Einstein's Berlin

Einstein's Berlin

In the Footsteps of a Genius

DIETER HOFFMANN

The Johns Hopkins University Press
Baltimore

Originally published as *Einsteins Berlin: Auf den Spuren eines Genies*
© WILEY-VCH Verlag GmbH & Co. KGaA, 2006

© 2013 The Johns Hopkins University Press
All rights reserved. Published 2013
Printed in the United States of America on acid-free paper
2 4 6 8 9 7 5 3 1

The Johns Hopkins University Press
2715 North Charles Street
Baltimore, Maryland 21218-4363
www.press.jhu.edu

Library of Congress Cataloging-in-Publication Data
Hoffmann, Dieter, 1948–
 [Einsteins Berlin. English]
Einstein's Berlin : in the footsteps of a genius / Dieter Hoffmann.
  pages cm
Translation of: Einsteins Berlin : auf den Spuren eines Genies. Weinheim :
 Wiley-VCH, 2006.
Includes bibliographical references and index.
ISBN 978-1-4214-1040-1 (pbk. : alk. paper) —
ISBN 1-4214-1040-0 (pbk. : alk. paper)
1. Einstein, Albert, 1879–1955. 2. Einstein, Albert, 1879–1955—Homes and
 haunts—Germany—Berlin. 3. Berlin (Germany)—
History—1918–1945. 4. Berlin (Germany)—Description and travel.
5. Physicists—Germany—Berlin—Biography. I. Title.
QC16.E5H64413 2013
530.092—dc23
 [B]   2012050892

A catalog record for this book is available from the British Library.

*Special discounts are available for bulk purchases of this book. For more information, please contact Special Sales at 410-516-6936 or
specialsales@press.jhu.edu.*

The Johns Hopkins University Press uses environmentally friendly book materials, including recycled text paper that is composed of at least 30 percent post-consumer waste, whenever possible.

CONTENTS

Foreword, by Walter Kohn vii
Preface 2013 ix
Preface xi

Einstein in Berlin 1

1 The Berlin Apartments 5

2 Einstein's Workplaces in Berlin 27

3 Homo Politicus 95

4 A Circle of Friends and Acquaintances 131

Bibliography 159
Illustration Credits 165
Name Index 167
Location and Street Index 171
Institution Index 173

FOREWORD

When I was born, in 1923, Einstein was forty-four years old. His greatest scientific works lay behind him. But his great personal influence was yet to unfold.

Einstein was a simple man, without any inclination to overestimate himself. Yet he grasped the very gist of science, human nature, the history of our time, and the great challenges of the future. Side issues and exaggerations were foreign to him.

I had the good fortune of making the acquaintance of his friend Toni Mendel, who lived in Hamilton, Ontario, during the 1940s. Toni was a wealthy, highly cultivated widow. She had emigrated in the 1930s from Berlin, where her home had been a lively social and intellectual center attended, apart from Einstein, also by the great Indian poet Tagore. (Her son-in-law, Dr. Bruno Mendel, is mentioned in the published conversations between Einstein and Tagore.) A deep friendship developed between Einstein and Toni Mendel. At the time when I and two or three other young people from Toronto occasionally visited her, she was regularly exchanging personal letters with Einstein. They wrote each other once or twice a year. Her reading of excerpts from these letters, full of personal warmth and originality as well as sparkling wit, were the highlights of our visits.

In 1952, I spent a few weeks at the Institute of Advanced Study in Princeton, and Toni Mendel kindly handed me a letter of introduction to Einstein. Our paths crossed from time to time, at the institute or outside, always with a simple, mutual greeting. But I did not feel I had the right to impose upon the invaluable time of that great man, even for just a few minutes, merely to introduce myself. Later I was ashamed of myself—he certainly would have been very pleased about the letter written by his former girlfriend from their Berlin years.

Both of them died a few years later. Unfortunately, Toni Mendel's daughter decided to burn Einstein's wonderful letters. But for me and the other young visitors at that time, they remained cherished memories whose spirit has stayed with me ever since.

When I became director of the new Institute for Theoretical Physics at the National Science Foundation in Santa Barbara, I was pleased that the inauguration took place on the centenary of the birth of this great man. His thoughts, words, and deeds were—and still are—the guiding star we follow, whether as scientists or simply as human beings. In this sense, I wish the present volume a wide circulation.

Walter Kohn

PREFACE 2013

No other physicists or even scientists have the worldwide recognition and admiration that Albert Einstein has. Thus it was not by chance that *Einstein's Berlin*—occasioned by the one hundredth anniversary in 2005 of Einstein's miracle year—has now been published in America. Although Berlin and America are geographically far apart, for Einstein both were central to his life and work. Just as an admirer of Einstein in Berlin or Germany might have a great interest in Einstein's life in Princeton, so might an admirer in Princeton or America have a keen interest in Einstein's life in Berlin. This translation is intended to satisfy Americans' interest in Einstein's life in Berlin and attempts to explore the complex relationship between Einstein and Berlin that is reflected in his work, his residences, and his "Berlin network." For the reader who is fond of traveling, this book may provide the stimulus to visit Berlin and to stroll in "the tracks of a genius" and can be a useful guide for such an adventure.

I would cordially like to thank everyone involved in the creation of this American edition. Thanks are particularly due to the Berlin institutions that generously funded the English translation: the Heraeus Foundation (Wilhelm und Else Heraeus-Stiftung), the Helmholtz Center Berlin (HZB für Materialien und Energie), the Physical Society in Berlin, and the Max Planck Institute for the History of Science. The last also offered me its rich material and intellectual resources and, as my home institution, provided me with the institutional setting for my Einstein research.

I feel particularly obliged to Walter Kohn, professor emeritus at the University of California in Santa Barbara and Nobel Prize winner in chemistry for 1998, whose reminiscences of Albert Einstein and Toni Mendel are included in the foreword to this book.

I am grateful to Ann Hentschel of Stuttgart for his careful translation and to

Mark Walker of Schenectady, New York, who was instrumental in putting this project on the right track.

Last but not least I thank the publisher, the Johns Hopkins University Press, and its helpful staff and collaborators, in particular, Bob Brugger, Melissa Solarz, and copyeditor George Roupe. I greatly appreciate their engaged cooperation.

PREFACE

> Now I understand the Berliners' self-satisfaction. You get so much outside stimulation that you don't feel your own hollowness so harshly as in a calmer little spot. (1, vol. 8, trans. p. 13)
>
> A. Einstein, 1914

Scientific knowledge is generally valid, apparently independent of the time and place in which it emerges. Galileo's law of falling bodies or Einstein's light quantum hypothesis, just like any other natural law or fundamental scientific principle, ought to be verifiable at any time and at any place. Even though the acquisition of knowledge in the (natural) sciences, however, has its arenas, whether we are talking about Greek antiquity, a Berlin school of physics, or even Silicon Valley as the birthplace of modern microelectronics. The general validity of scientific knowledge and the dominance of the temporal in the history of science—as in historical representations generally—allow the space, the location, to fade almost entirely into the background and, at best, allow it to become an anonymous stage in the global theater of the history of science.

Thinking about space and time (together) in a new way was one of Einstein's great achievements, along with his theory of relativity, which provided a new basis for modern physics. Why shouldn't the historian of science also follow Einstein in joining together time and space in an entirely concrete sense and track down the scenes of the action? Existing biographies portray in varying degrees of detail the time line on which Einstein lived and worked. Information about the concrete spaces in which he moved is much more scarce. This is what I explain in the present book.

The focus is Berlin, where Einstein lived and worked for almost two decades. During this period, between 1914 and 1932, the worldwide fame and mythical figure of Einstein were formed. This period also marks the high point of his recognition in science and society.

Despite the central role that Berlin played in his life, the city did not appeal to Einstein at all. He repudiated the militant and authoritarian spirit for which the Prussian capital and Wilhelminian Germany as a whole stood. Any larger city or metropolis generally did not go down well with Einstein, and he kept a wary distance from them.

Except for a one-year intermezzo in Prague, Berlin remained the only major city in which Einstein chose to settle down. The busy streets and the superficiality and distractions of city life were, in his opinion, detrimental to deep scientific contemplation. Other great contemporaries—like the playwright Bertolt Brecht—took a different view and even adopted the city as the heart of their creative and working lives. Einstein decided to come and live in Berlin in 1914 because of the ideal working conditions for him there. The aura of great scientists and intellectuals had a magical attraction. Particularly in physics, a unique atmosphere of creativity prevailed from 1870 to 1933, which closely linked general progress with the activities of Berlin's physicists. Einstein was well aware of this and considered it an honor and a challenge to be allowed to join this scientific elite. He certainly valued the "fine human ties" he was able to make there. His disappointment was great when, almost twenty years later, he was driven away from this scientific and intellectual center, and this same scientific elite showed absolutely no solidarity, rushing instead to the service of the ruling National Socialists.

These general historical conditions in science and politics obviously set the framework of this book, but rather than being elaborated further, they will serve to interconnect the localities of Einstein's life and work in Berlin. His places of residence and employment are described, along with the venues of his various lecturing and political engagements or rendezvous with his personal friends and acquaintances. So this book is neither a biography of the genius scholar nor an exhaustive documentation of the Berlin segment of his life nor even a description of his importance for Berlin then and now. More about that can perhaps better be gathered from the pertinent biographical literature (see references in the bibliography). Nevertheless, this book aims to be more than a historical guidebook for tourists and others interested in the important Einstein sites of the city. By following the concrete traces of this scientific revolutionary and citizen of the world in the microcosm Berlin, this book aims to convey how his life and work were linked with the scientific and social life of the city: how strongly Einstein was bound within a "Berlin network." The reader will furthermore learn of what scientific and intellectual culture Berlin lost through the crimes of National Socialism and the destruction of World War II. The historical photographs also document the material losses and the altered face of the city today. I hope that following in Einstein's footsteps in exploration of the city of Berlin will also arouse the reader's curiosity and enterprising spirit to continue following the traces of Berlin's glamorous scientific past and approach the history of science in an entirely different way.

In the hope of having documented Einstein faithfully, I thank all those who contributed substantially with their suggestions and criticism in the production of this little book. Besides my gratitude to numerous local historians, libraries, and archives, I would specifically like to name my colleague Giuseppe Castagnetti, who readily shared with me his extensive findings on Einstein and generously helped clarify numerous details—as did Barbara Wolff from the Einstein Archive at the Hebrew University of Jerusalem. I also most heartily thank Hubert Laitko (Berlin) and my wife, Reina Hoffmann, for their critical review of the German manuscript. The linguistic quality of the book ultimately benefited by it. Edith Hirte, Konrad Reissmann, and Heinz Reddner helped me with the time-consuming search for illustrations and acquisition of the necessary prints. The Max Planck Institute for the History of Science must also be acknowledged; the present book gained substantially by the institute's competence and resources on the subject of Albert Einstein. My thanks go to the publisher, Wiley-VCH, and particularly to Mrs. Esther Dörring, for her patient and constructive collaboration. The W. E. Heraeus-Stiftung generously supported the publication of the German edition of this book with a grant for the printing costs.

Einstein in Berlin

> The Berlin gentlemen are speculating with me as if I were a prize-laying hen. (20, p. 30)

"At Easter I'm going to Berlin as an Academy person without any duties, somewhat like a live mummy. I'm looking forward to this difficult profession" (1, vol. 5, p. 538), wrote Albert Einstein to his friend Jakob Laub in summer 1913. As a "paid genius," Einstein worked in the city for almost two decades—until the Nazis forced him to emigrate as a figurehead of so-called Jewish science and as an intellectual representative of the hated Weimar Republic.

At that time Berlin was a center of intellectual life and, in particular, an internationally acclaimed stronghold of scientific research—the city could boast brilliant scholars and world-famous research institutions, specifically in physics. Through them, the development of physics as a whole was firmly tied to physics research in Berlin. When Einstein made the personal acquaintance of leading Berlin physicists in the fall of 1911 at the Solvay Conference—a summit meeting of leading physicists of the day—and formed some friendships on that occasion, efforts were redoubled to win over the new genius, a shooting star in the skies of physics. Then, as today, brilliant minds simply attracted other brilliant minds. But it was by no means a matter of course that Einstein would accept the advances of the Berlin physicists, and he wavered a long time before reaching his decision. He eyed the advancing "Berlin adventure . . . not without a certain unease" (1, vol. 5, trans. p. 373). His relationship with Germany and specifically his attitude toward the Prussian mentality of militant authoritarianism were highly ambivalent. As a sixteen-year-old, he had fled from it and had never felt the least inclination to return.

It was not just the prospect of becoming a member of a scientific community that represented the world's elite in the field of physics that brought Einstein to the banks of the Spree in anticipation of important stimuli for his own work.

The attractions of the gentle sex, in the form of his cousin Elsa Einstein, also helped decide the matter. During his visits to Berlin in 1912 and 1913, the happy reunion of the two cousins evolved into a relationship of mutual respect and love. After one of his visits, a still quite starry-eyed Einstein recalled a walk they had taken along the shores of the Wannsee, with the exclamation "If only I could repeat it!" (1, vol. 5, p. 456). Two years later he moved to Berlin, and in February 1919 they were married.

It had not been a simple matter for the physicists in Berlin to arrange for the appropriate conditions for Einstein's appointment in the German capital. Their first attempt to create a suitable position for him fell through. Finally, in the spring of 1913 a bow could be tied on an appointment package that was not only attractive to Einstein but also acceptable to all parties. As his main occupation, Einstein was invited to accept membership in the Prussian Academy of Sciences, as a "paid genius," so to speak. The reasoning was that he would be able to concentrate exclusively on his research interests. This was different from the situation of his fellow academicians, whose main occupations were as university professors or directors or as employees of unaffiliated research institutions. In accordance with the academy's character as a scholarly society, their duties were restricted to meeting there regularly to exchange ideas about their research results. To make the position for Einstein at the academy attractive not just intellectually but also financially, an unbureaucratic solution was found. He was offered the maximum salary of a German university professor, but the academy, or the Prussian state, could not come up with the annual sum of about 12,000 marks on its own. So a Berlin banker, Leopold Koppel, who had already distinguished himself as a generous patron of the sciences at the founding of the Kaiser Wilhelm Society, was approached. Koppel agreed to help finance Einstein's academy position. A professorship at the University of Berlin was also linked to Einstein's academy membership. This granted him all the rights of a university teacher without the obligation to give regular lectures. Einstein was also promised the directorship of a new, yet to be founded Kaiser Wilhelm Institute of Physics.

What Einstein was hoping to gain by his move to Berlin was access to scientific results from the latest research; the Berlin physicists, for their part, were hoping that Einstein's appointment would allow them to enlist him in their efforts to solve the many new fundamental problems in physics that the development of quantum theory had laid bare. They were particularly hoping for a new theory of matter that would promote an integration of physics with chemistry. Neither of these mutual expectations materialized, however. Einstein focused

Fig 0.1. Albert Einstein with Berlin colleagues, Dahlem, 1921

his concentration entirely on his general theory of relativity. When he succeeded in completing it in the fall of 1915, he promptly reported on this triumph at the meetings of the academy. Yet the hope of strengthening the institutional spectrum of Berlin physics by the founding of a Kaiser Wilhelm Institute of Physics was not fulfilled. The institute essentially remained a "one-man operation" for the distribution of research grants, and in the 1920s Einstein withdrew from these duties as well.

Despite all these dashed hopes, Einstein's scientific achievements in Berlin remained impressive. Besides the finished general theory of relativity, he proved the gyromagnetic effect and contributed to the quantum theory of radiation and the formulation of the Bose-Einstein statistics. The spectacular confirmation of the general theory of relativity transformed him in the 1920s into a public figure—the first star scientist of the dawning age of mass media. The Berlin years thus mark the high point of Einstein's scientific and social prestige. But they were also a time of growing political and anti-Semitic aggression against Einstein personally and against his work. During the early years of that decade he even became the target of assassination threats, yet Einstein still felt so firmly a part of the Berlin scientific community that foreign universities were unable to coax him away. In September 1920 he confessed to the Prussian minister of culture: "Berlin is the place in which I am most deeply rooted through personal and professional ties. I would follow a call outside of the country only in the case that external circumstances forced me to do so" (21, p. 204; 1, vol. 10, trans. p. 263).

Such a situation arose when the National Socialists grabbed the reins of power in January 1933. Einstein's overly compliant fellow academicians bent under the pressure of a few Nazis among them and voted to expel him, but Einstein had already sent his official resignation by the end of March. But he was unhappy to do so—as his letter to the academy explains—"because of the intellectual stimulation and the fine human relationships which I have enjoyed throughout this long period as a member and have always valued highly" (21, vol. 1, p. 246). Einstein's departure from the academy was the beginning of an unprecedented wave of expulsions of scientists and artists from Germany. A unique creative atmosphere—that imponderable and yet very real quality of intellectual fecundity that transformed Germany, and specifically Berlin of the first decades of the twentieth century, into a world center for science—was allowed to fall apart and important elements of it to be destroyed within a short period of time. The consequences are still perceptible today, for although scientific institutions can be rebuilt and new talent can be found, that atmosphere of high intellectual activity with its irresistible attractive pull that Einstein's work of some twenty years signified was the work of many generations and a heritage well worth safeguarding. Difficult as it was to build up, it was all too easily lost.

CHAPTER ONE

The Berlin Apartments

Site 1. The first Berlin apartment
Ehrenbergstrasse 33
14195 Berlin (Dahlem)

subway stop Thielplatz (U1) or
met. train sta. Lichterfelde West (S1),
from either, 10 min. by foot

Einstein's first apartment was located in Dahlem, at Ehrenbergstrasse no. 33. His choice to move into this expanding Berlin suburb rather than into the inner city was not by accident. As we already know, the position offered him included the prospect of becoming the founding director of a new Kaiser Wilhelm Institute devoted to physics, as well as membership in the Prussian Academy and a professorship at the University of Berlin. The Kaiser Wilhelm Society, recently founded in 1911, was located in Dahlem. Other Kaiser Wilhelm Institutes for chemistry and physical chemistry and electrochemistry had also just been opened with due pomp and ceremony in the fall of 1912. More were soon to follow.

After accepting the appointment to the Berlin academy in December 1913, Einstein sent his wife, Mileva, out in advance to look for an apartment. She was helped by the director of the Kaiser Wilhelm Institute of Physical Chemistry, Fritz Haber, and his wife, Clara. It was not long before she found a suitable apartment—not far from Haber's institute, in which Einstein would set up his first office—and he signed the lease as of April 1, 1914. The building, at the corner of Ehrenbergstrasse and Rudeloffweg, belonged to the master painter Johann Nikleniewicz from Lichterfelde and had been newly built in 1910. "The new landlord is very decent. He's having everything renovated nicely," Einstein reported to his wife in April 1914 (1, vol. 8, trans. p. 9).

Einstein had arrived in Berlin on March 24, 1914, initially without his wife and children, who were still undergoing a health treatment in Ticino, Switzerland. Because the apartment was still under renovation and not all the furniture had arrived in Berlin, Einstein first lived with his uncle Jakob Koch on

Fig 1.1. The building at Ehrenbergstrasse no. 33, fall 2004

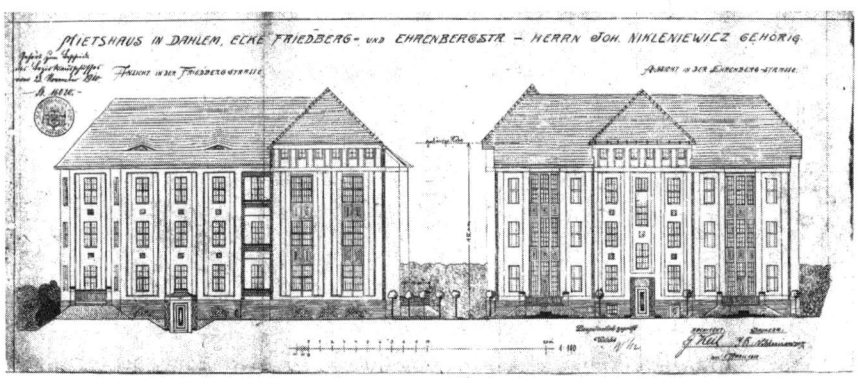

Fig 1.2. Drawing of the façades of the building at Ehrenbergstrasse no. 33, 1910

Wilmersdorfer Strasse. Einstein's mother, née Pauline Koch, was keeping house there. In mid-April the apartment on Ehrenbergstrasse was finally ready and furnished (on which floor it was located is not known), so Mileva and the boys were able to resettle in Berlin. But their life there together as a happy family was short. Smoldering conflicts between the couple flared up again—Einstein's affair with his cousin Elsa and the tensions between Mileva and her in-laws, particularly Einstein's mother, only exacerbated the situation.

Fig. 1.3. Mileva Einstein with her sons Eduard and Hans Albert, 1914

In June the quarreling culminated in Mileva leaving the apartment at the end of the month, taking the children with her. They found refuge in Haber's spacious villa at Hittorfstrasse no. 24, which he used for professional purposes. When all mediation attempts by the Habers and others had failed, Mileva left permanently with the children for Zurich at the end of July. Einstein described the parting scene to his cousin:

> The last battle has been fought. Yesterday my wife left for good with the children. I was at the railway station and gave them a last kiss. I cried yesterday, bawled like a little boy yesterday afternoon and yesterday evening after they had gone. Haber accompanied me to the station (9 o'clock) and then spent the evening with me. Without him I would not have managed to do it. (1, vol. 8, trans. p. 37)

That was the end of Einstein's first marriage; the divorce was finalized in February 1919.

Einstein could not stand living alone in the roomy apartment on Ehrenbergstrasse for long. In late 1914, probably in November, he decided to move closer into town, near Fehrbelliner Platz in Wilmersdorf—not very far from his cousin's residence.

Site 2. The second Berlin apartment
Wittelsbacherstrasse 13
10707 Berlin (Wilmersdorf)

subway stop Konstanzer Strasse (U7),
from there 3 min. by foot

In his second Berlin apartment Einstein lived the life of a bachelor; as he wrote his trusted friend Heinrich Zangger, he was "living a very secluded and yet not lonely life, thanks to the loving care of my cousin, who had drawn me to Berlin in the first place, of course" (1, vol. 8, trans. p. 110). He must have been among the first renters of the building, since it was only completed in 1914. Its architect and owner was Franz Abbe. Einstein is listed in the Berlin directory for 1915 as inhabiting the third story. According to the building blueprint, it contained a spacious seven-room apartment and a smaller three-room unit. Whether he moved into the larger apartment, as befitted the status of an academician and professor, is not clear. The above quote and Einstein's modest attitude toward material things in general would lead one to think that the smaller unit would have entirely sufficed for him. A letter to his older son from January 1915 supports this conjecture. There Einstein relates how he was now "living in the city, near Fehrbelliner Platz. There I have a little flat" (1, vol. 8, trans. p. 64). His apartment also had a telephone connection, which was not uncommon in that wealthy middle-class neighborhood. By then telephones were already a common convenience for people of Einstein's standing; just a decade before, they were not at all. When Walther Nernst moved to Berlin in 1905, for example, he was considered technologically avant-garde to possess his own telephone among very few other professors in the city.

Otherwise we know relatively little about Einstein's exact living circumstances. According to the report of one visitor, Einstein seems to have been "lodging" in a quite empty apartment (51, p. 25); when a student came by one day to ask for a publication, Einstein searched about for it for a long time and "wondered where it could possibly be, complaining about his untidiness and forgetfulness, and could not find it. We went from room to room and stood mystified in front of the bookstands. He has a quite bare apartment and seems to live alone there and without a housekeeper" (15, p. 401). To his elder son Einstein reported that he "usually work[ed] all day" there, sometimes even cooking his own lunch. In a letter from this period to his trusted friend Zangger we read: "My human

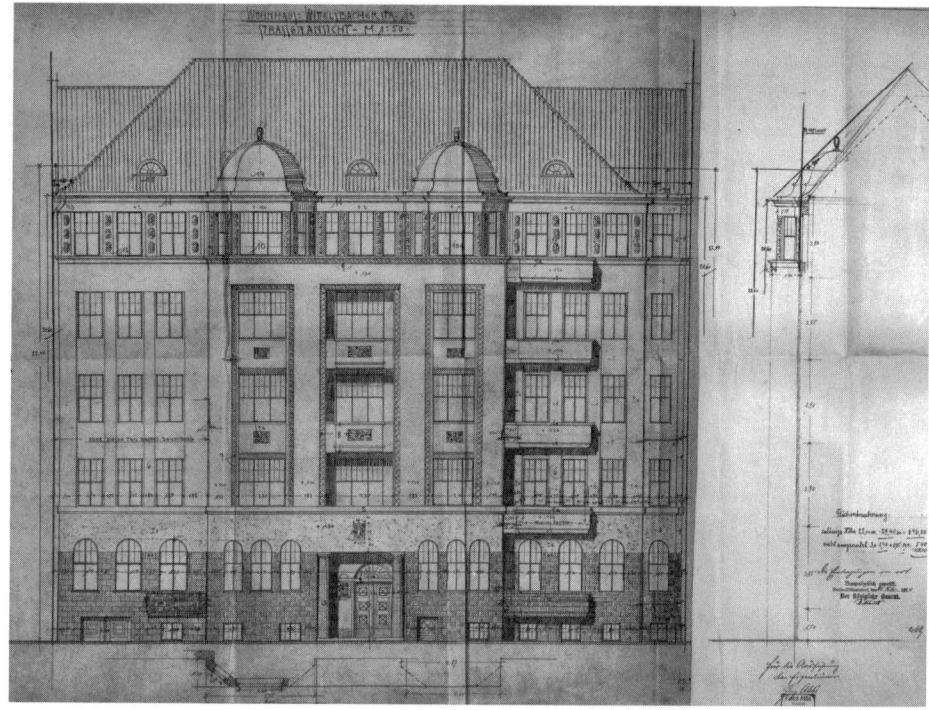

Fig. 1.4. Façade drawing of the building at Wittelsbacherstrasse no. 13, 1913

and professional contacts are few but very harmonious and rewarding, my public life withdrawn and simple. I must say that to me I seem one of the happiest of persons" (1, vol. 8, trans. p. 88). Elsewhere he describes his life as "isolated by mentality and outlook on life" (1, vol. 8, trans. p. 300).

In this self-imposed isolation Einstein was able to concentrate entirely on his work on the general theory of relativity and to bring it to successful completion in the fall of 1915. This scientific achievement, with its quite considerable psychological and physical stresses under the bad living conditions of wartime, had its price, though. At the beginning of 1917 Einstein fell seriously ill, first with jaundice, then stomach ulcers and gall stones; later other health problems were added to the list. His doctor put him on a strict diet, but he only managed to keep to it during the winter of 1916–17 and the following lean months thanks to regular "family packages" of food from southern Germany and Switzerland, along with the selfless attention of his cousin and lover Elsa Einstein. It was also Elsa who convinced him to move out of his bachelor's flat that summer. In December Einstein reported to Zangger: "I have gained ca. 4 pounds since the summer,

Fig. 1.5. Albert Einstein, probably 1916

thanks to Elsa's good care. She cooks everything for me herself, since this has proved to be necessary. This is possible because I live in the apartment next to hers (for the interim)" (1, vol. 8, trans. p. 412).

Site 3. The third Berlin apartment
Haberlandstrasse 5
10779 Berlin (Schöneberg)

subway stop Bayerischer Platz (U4 or U7),
from there 3 min. by foot

Einstein's move in late summer 1917, probably about September 1, 1917, into Haberlandstrasse was surely not just because of the need for his cousin to nurse his long illness. The personal situations of both of them had developed enough in recent years to make a more permanent commitment between them conceivable. One obstacle was that Einstein was still married and Mileva was adverse to agreeing to a divorce. Einstein, too, seems for a long while to have rather been interested in maintaining a "meal-ticket relationship" over spousal companionship. The parenthetical "for the interim" in the letter quoted above suggests this, as does his remark to a friend, Michele Besso, that "the peace and tranquility does me tremendous good, and no less so the extremely agreeable, really fine relationship with my cousin, the permanent nature of which is guaranteed by a renunciation of marriage" (1, vol. 8, trans. p. 68). His severe illness and Elsa's self-sacrificing care evidently led to a change of heart on this question—or shall we say, it created a new state of affairs, since the driving force behind this move had apparently been Elsa. "My address is thus 5 Haberlandstrasse. The move seems to have already been completed" (1, vol. 8, trans. p. 372), Einstein wrote his friend Michele Besso at the beginning of September while away on a trip. In apparent keeping with bourgeois conventions, Einstein initially moved into an apartment separate, yet "connected" to that of his cousin. It was located on the fourth story and was, in Einstein's words, "spacious and comfortable" (1, vol. 8, trans. p. 374).

Fig. 1.6. Haberlandstrasse 5 apartment building, 1920s

Haberlandstrasse is a part of what was called the Bavarian quarter, situated at the boundary between the Berlin districts of Wilmersdorf and Schöneberg. At that time it was a well-to-do residential area where many prominent Jewish intellectuals lived. Einstein's neighbors included the writers Else Lasker-Schüler and Artur Landsberger, the theater critic Alfred Kerr, the mathematician and world champion chess player Emanuel Lasker, as well as Rabbi Leo Baeck. Because of its high percentage of Jewish inhabitants, this quarter was also occasionally called "Jewish Switzerland." The Third Reich transformed it completely. Many of its residents were forced into exile; countless numbers of them were murdered in the extermination camps. The street itself was renamed Nördlingerstrasse in 1938 because its original namesake had also been Jewish; it got its old name back again only in the 1990s. Large swathes of Haberlandstrasse, including no. 5, were destroyed in the bombing of World War II, and completely new buildings were erected there later.

The building at Haberlandstrasse no. 5 was on a corner. In addition to the main entrance with its typical stairway, doorkeeper, and elevator, there was also a separate access for servants and other employees from Aschaffenburger Strasse no. 17. The building had been completed by the architect Otto Eisfelder in 1907–8, who shared ownership of it with Elise Pulsack. Elsa Einstein's parents, Rudolf and Fanny Einstein, must have been among their first tenants. They first lived in an apartment on the fourth story, moving down to the third in 1912. In the 1920s they eventually descended to the first-story apartment. Their daughter Elsa also came to live in that building following her divorce. She brought along her own two girls, Ilse and Margot, probably first staying with the elders, later taking over their apartment (around 1912). Einstein's apartment "for the interim" was on the fourth story. It is unclear when exactly he vacated it. In any event, Einstein officially moved into Elsa's apartment after their wedding on June 2, 1919. It is interesting that at the civil registry Einstein's address is listed as a guesthouse at Uhlandstrasse nos. 113/114. A passage from Einstein's letter to Elsa written in the summer of 1914—that is, directly after his separation from Mileva—indicates how careful one had to be about adhering to the social conventions of the time:

> Haber has impressed upon me that we must be dreadfully careful so that we, i.e., you do not become the subject of idle gossip. Do not go out alone! Haber will inform Planck so that my nearest ~~relatives~~ colleagues do not first hear about the matter from rumors. You will have to perform wonders of tact and restraint so that you

are not looked upon as a kind of murderess; appearances are very much against us. (1, vol. 8, trans. pp. 37–38)

Acting "very saintly during this time" (1, vol. 8, trans. p. 36) did not exclude going on vacation together by the Baltic in the following summers—albeit in the guise of family excursions: they were cousins, after all.

The Einsteins' apartment was spacious even from the point of view of his contemporaries. That is how Einstein's biographer and colleague Philipp Frank also described it. To Charlie Chaplin, however, it seemed "modest and small": "You could have found the same apartment in the Bronx, too, a living room that also served as a dining room. Old, worn carpets were lying on the floors. The black grand piano was the most valuable piece of furniture" (78, p. 327). There were seven rooms in the apartment as well as a side room. Konrad Wachsmann, the architect of Einstein's summerhouse, described the apartment in 1979 as follows:

> If you entered the hallway, Einstein's bedroom was on the left, behind that was the library and the salon in which the grand piano stood. From the salon you could pass through a sliding door on the right-hand side into the dining room. Straight ahead was another door through which you came to a small hallway and from there to the bathroom. Elsa Einstein's and her daughter Margot's bedrooms also issued into this hallway. I do not know which room the daughter Ilse inhabited because she was already married. Behind the kitchen there were some more rooms for the staff. But I never saw them. (22, p. 13)

Many famous visitors made their way to Einstein's apartment. The list ranges from men of the arts like Gerhart Hauptmann and, as already mentioned, Charlie Chaplin, to Einstein's patron, the banker Leopold Koppel, and the politician Walther Rathenau. The writer and diplomat Count Harry Kessler recalled a social event at the Einsteins' one evening in the spring of 1922:

> Ate in the evening at Albert Einstein's. Quiet, pretty apartment in the west of Berlin . . . , a slightly too copious and mass-produced diner, lending this truly dear, almost childish couple a certain naïveté. . . . Some sort of radiant kindness and simplicity transported even this typical Berlin society from the conventional and illuminated them in an almost patriarchical, fabulous aura. (87, p. 278)

Aside from these rather rare social evenings, there were much more frequent teas or musical evenings in the Einstein home for a handful of friends or col-

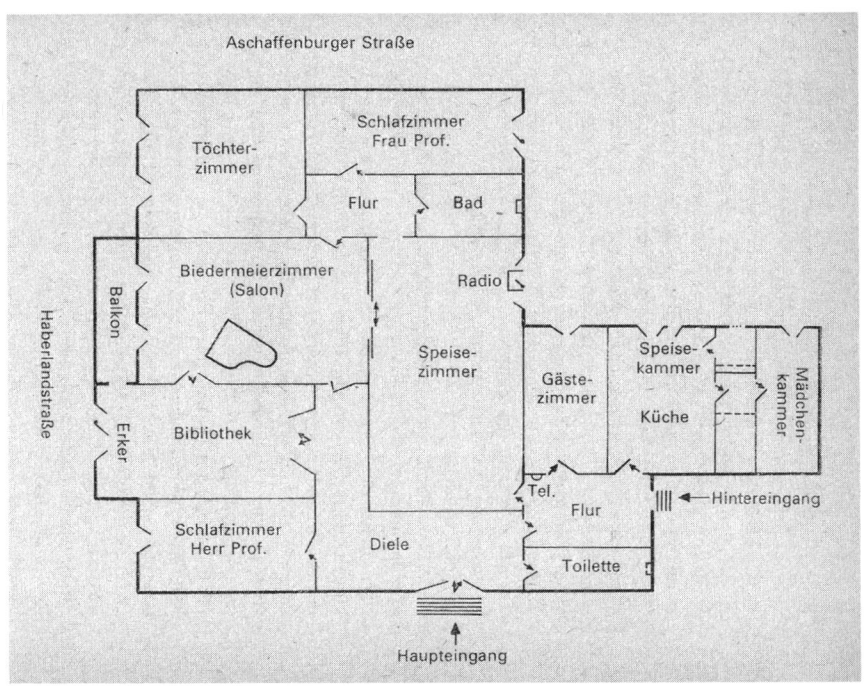

Fig. 1.7. Floor plan of the apartment of Albert and Elsa Einstein

Fig. 1.8. Music-making at the home of the Einsteins, late 1920s. *From left to right:* the cellist Francesco von Mendelssohn, the pianist Bruno Esser, and Albert Einstein.

leagues. From time to time invitations to listen to reports by Einstein about his travels or other matters of interest were sent out.

His colleagues and coworkers were more frequent guests at Haberlandstrasse. They were usually received in what was called the tower room. This room was the product of a remodeling of the attic in 1922 so that Einstein could have a place to work without disturbance and settle his official business from home. He did not have an office at the academy for his personal use, even as director of the Kaiser Wilhelm Institute of Physics. So Haberlandstrasse had to serve as the official business address. The hideaway under the roof consisted of three rooms: the actual office of about seventeen square meters, a storage area for books, and another room. Because no official permit had been applied for before the attic renovation was done, the Municipal Building Inspectors Office ordered Einstein to move his things out again for structural and hygienic reasons. Einstein responded with an "application for dispensation" addressed to the Berlin chief of police:

> I had to have the space renovated and a separate access to it built from the stairway in order to be able to attend to my studies away from my apartment, in which I have been disturbed far too often. I disbursed a substantial sum for this, which is considerable for an averagely compensated civil servant/university professor.... The room is supposed to be used only by me personally, not, however, by others. Any eventual hygienic defects could only affect me personally; I know, however, that most people in Berlin have to work in much less favorable accommodations.... As a reputable scholar and teacher at the university, I have moral claim to special treatment with regard to my application concerning my study. (47, trans. p. 222)

The building office evidently could not reject Einstein's request, owing to his great popularity and fame, so his illegal construction work was eventually given official sanction after the fact.

The office, with its desk, armchair, and bookshelves covering the full length of the wall, a small table, an easy chair and stool, was perfectly suited to Einstein's working requirements. A small telescope, with which Einstein could observe the sky and which visitors and occasionally neighbors inspected, completed the furnishings. The walls were hung with portraits of the physicists Michael Faraday and James Clerk Maxwell as well as the philosopher Arthur Schopenhauer. The main apartment was somewhat of a contrast to this Spartan arrangement. Its decor in the Wilhelmine style could only be attributed to Elsa's taste in tune with the bourgeois norms of the day. It reflects the ideal of beauty and comfort held

Fig. 1.9. Albert Einstein in his tower room on Haberlandstrasse

by the educated and well-to-do *Bürger*. In Konrad Wachsmann's recollection Einstein himself always looked "as if he had wandered into these rooms entirely by chance and now has to live there because he doesn't know where the exit is" (22, p. 141). This impression is confirmed by Philipp Frank, who was Einstein's successor at Prague and a frequent visitor in Berlin. He recalled that, although his host lived "in the midst of fine furniture, carpets, and pictures," one constantly had the feeling that "Einstein forever remained a stranger in such a 'bourgeois' household" (26, p. 154). The lost-looking figure in the photograph below also corroborates these observations.

When Einstein and his wife left on a research tour to the United States on December 6, 1932, they were fated never to see their home again. When the National Socialists took over power, the SA storm troopers conducted a number of searches there in the spring of 1933, at which time some of their belongings were taken away. But most of their personal effects—including the piano and, most important, Einstein's library and files—could be saved. They were packed up that summer and, with the help of the French embassy, arrived in Princeton by special delivery via Paris and became the foundation of Einstein's new home in America.

Fig. 1.10. Albert and Elsa in the salon

Site 4. The summerhouse
Am Waldrand 15–17
14548 Schwielowsee (Caputh)

By met. train (S7) or regional express train
  to Potsdam main station
then by bus (line 607) toward Caputh/Ferch
to the Caputh-Kinderheim stop,
from there 10 min. by foot uphill

"Deep contemplation does not thrive near activity. That's why metropolitan life is not appropriate for researchers and students" (37, p. 272), declared Einstein in 1924, leaving no room for doubt about his lifelong wariness of big cities. We know that in Munich, where he spent his childhood and youth, he grew up in a virtually rural setting, acquiring a strong sense of nature at an early age. Aside from the period in Prague, his academic career was unconnected to any major city. So it is not surprising that once in Berlin, Einstein would put much effort into creating a country refuge for himself. This was not unusual for Berliners. Those who could afford such a thing—like his colleague Walther Nernst—had a country estate on the outskirts, and it was common practice even for laborers to use a gardening plot or hut for recreation.

The Einsteins also had one in the early 1920s that Elsa rented out for the family. It was located in the gardening settlement in Boxfelde, Spandau, at Burgunderweg no. 3. In his letters Einstein jocularly referred to the hut as his "Spandau castle." He vacationed there in the summer of 1922 with his two sons: "The boys are here and are staying in my Spandau castle. I commute back and forth like this between the city apartment and the castle which, unlike my yacht, proves to be watertight" (31, p. 177).

We know very little about "Spandau castle" and how its prominent tenant used it, but this gardener's paradise was probably not Einstein's idea of the perfect idyll. Such a Berlin lot simply wasn't intended purely for leisure and relaxation. Then, as now, one was expected to tend it properly. If anecdotes and recollections from that time may be believed, Einstein seems not to have done so to the satisfaction of the other lot keepers, and his lease for "Spandau castle" was terminated after a short time, probably by mutual agreement.

The long-cherished wish for a countryside idyll was eventually fulfilled in an entirely different location. At the beginning of 1929 the Municipal Council

Fig. 1.11. "Spandau castle" (probably), 1921

of Berlin came up with the idea of signing over a piece of land by the water to their world-famous citizen in honor of his fiftieth birthday. This generous plan became a political disaster for the city, however, when it turned out that the city had no authority over the house and grounds initially envisioned for the purpose in Kladow nor over two other pieces of real estate that were subsequently offered to him. To correct this political fiasco Einstein was asked to look around for a suitable property himself. Elsa then found one in the community of Caputh near Potsdam, so the Municipal Council was able to report in a submission for resolution on April 24, 1929:

> The Municipal Council has decided to present to the greatest scholar of our century, our fellow citizen Professor Einstein, a parcel of land on the Havel on his 50th birthday. The discussions conducted with Mrs. Einstein in this regard yielded a description of a parcel of land that meets all the great scholar's demands regarding tranquility, pretty situation, access to the Havel for sailing as a sport, close

connections with public means of transportation and convenience for deliverers. The land is located in Caputh, Waldstrasse nos. 7/8, has a fine far view across the Havel, and is described as excellently suited to Professor Einstein's purposes. We ... shall have a raised foundation added to the gently rising slope to improve the view and also lay gardens. The building itself will be erected by Professor Einstein. (28, p. 257)

Nothing came of the honorary gift in the end, though, because the Assembly of City Representatives became hopelessly locked in debate about it when conservative and nationalistic elements tried to sway opinion against Einstein. To settle the problem, Einstein told the council and the chief mayor that he would have to decline any gift from the city of Berlin, because he could not reconcile the continual delays with the nature of a gift. Besides, life was too short for such honorary awards. The matter was finally closed, for the city at least, uncomfortable though it was, but not for Einstein. He decided to come forward as the purchaser himself and soon afterward, on June 21, 1929, a permit was issued him by the local building authority to construct a summerhouse on the property.

Fig. 1.12. View of the summerhouse

Einstein assigned the job to the twenty-eight-year old budding architect Konrad Wachsmann. Having read about the proposed honorary gift by the Berlin Magistrat in the papers, Wachsmann had simply rung at the Einsteins' door on Haberlandstrasse and introduced himself, proposing to Einstein (or his wife) that he be entrusted with the construction of their summerhouse. At that time Wachsmann was still employed as an architect for one of the largest wood construction companies in Europe, Christoph & Umack in Niesky, Oberlausitz. So it was not coincidental that Wachsmann would suggest a wooden house made of prefabricated components. The drawings he submitted a few days later—the house was supposed to be a simple, straightforward, and functional product of modern architecture—were so convincing that he received the commission. Having shown only a peripheral interest in the decor of the Haberlandstrasse apartment, Einstein wanted to know absolutely everything about the Caputh project. According to Wachsmann, he was full of questions about

> how the house was supposed to be built and what should be put up where. According to his conceptions, the house was supposed to be made of brown-stained wood, the windows had to be white, equipped with a thin iron parapet and white wooden window shutters. Like typical French windows, which extend from floor to ceiling. Furthermore, in keeping with the owner's conceptions, there should be enough open and wind-sheltered terraces to be able to live as much out of doors as possible. (22, p. 127)

Although Wachsmann altered Einstein's initial conception to conform with specific requirements, the final design stayed essentially true to the basic idea. The construction of Einstein's summerhouse became the beginning of a successful career for the architect.

By July 17 the foundations and the basement were up, and in the following weeks the wooden structure was erected. In October the building inspections could be performed and the move begun. The Einstein family had meanwhile already found lodging in Caputh in a house on Potsdamer Strasse, which allowed the Einsteins to supervise the construction and still enjoy the summer in the countryside, and the professor was able to do his beloved sailing.

Einstein's summerhouse is still strikingly modern in style and ambiance. The house was built of Oregon and Galician pinewood and overlooks a broad view of the surrounding countryside including the River Havel and Templiner See to the southwest. What was called the garden room on the ground floor forms the center of the house: a room of about forty square meters with an open hearth and two large folding doors opening out onto the roofed garden terrace.

The ground floor otherwise accommodated a small, very functional kitchen with a service hatch to the garden room, a bathroom, and water closet, as well as bedrooms for Elsa and Albert Einstein.

Einstein's bedroom was also his study. Besides a berth, it also contained a desk and one bookcase that had not been part of the original plan. The window of this room offered an unobstructed view of the garden and the Havel landscape. Three more, smaller bedrooms were situated on the second level of the house for guests or employees. From there one had access to the roof terrace, which could also be reached by a separate stairway from the garden. The main entrance is at the back of the house on the ground floor toward the woods. The official address at that time was Waldstrasse nos. 5/7.

Konrad Wachsmann was responsible not just for the construction but also for the interior design of the entire house. His plan was to decorate the house with furniture by the Bauhaus architect and designer Marcel Breuer, but his initial sketches were not well received. Einstein supposedly apodictically declared: "I

Fig. 1.13. Einstein's study and bedroom in Caputh

really don't want to sit on furniture that constantly reminds me of a machine shop or an operation room! No, I don't like these things at all" (22, p. 130). Here, as in other instances, Einstein reveals a rather conventional and antimodern taste, so Wachsmann was forced to furnish the house with furniture of his client's liking—such as Einstein's desk—and other pieces Wachsmann had chosen had to be dispensed with in Berlin or stowed away in the attic. Wachsmann did manage to incorporate one modern detail in the interior design, though: the lamps from the Bauhaus assortment apparently did appeal to Einstein. "The little house in Caputh is a bankruptcy, but a very nice one, not to mention the sailboat, which high finance gave me!" (22, p. 313). This statement in a letter to his sister Maja from the fall of 1929 is typical of how he described his new summer abode. The land purchase and the construction costs had used up most of the family's savings. But the effort expended on it was compensated by "the sailboat, the view into the distance, solitary autumnal walks, the relative quiet; it is a paradise" (22, p. 312). Caputh became Einstein's place of retreat, where he could take long walks in the woods and go sailing as he so much liked to do. But best of all, he could concentrate on his work away from the bustle of city society. Elsa reported to Hedi Born in September 1930, "Albert slaves away like never before, he radiates and shines. He's thought up the most magnificent theory.... If only it stays true!!!" (16, p. 332).

The presence of non–family members was a part of the working atmosphere at Caputh. Einstein's collaborator at the time, Walther Mayer, stayed there for many weeks, and the secretary Helen Dukas also came out regularly to deal with the correspondence and settle other business. Close friends and colleagues like Max von Laue, Max Planck, Paul Ehrenfest, and Erwin Schrödinger also sought out the Brandenburg idyll to exchange thoughts with Einstein about scientific matters or were invited to go out on the sailboat or on a hike in the woods. The Caputh domicile became the meeting place of preference for the family. Albert Einstein's stepdaughters Ilse and Margot stayed in the guestrooms on the upper floor of the house during the summer months, and his sister Maja was also an occasional visitor. His sons Eduard and Hans Albert also came to stay at Caputh more than once. The elder of the two arrived one time early in the summer of 1932 on a motorbike with his two-year-old son Bernhard Caesar. The house also attracted prominent visitors like Heinrich Mann, Arnold Zweig, Max Liebermann, and the young Anna Seghers, who came to Caputh to invite Einstein to give a talk in the Marxist School for Workers. The visit of the Indian Nobel laureate poet and philosopher Rabindranath Tagore in the summer 1930 was particularly spectacular. An article about it even appeared in the papers.

Fig. 1.14. Einstein with his son Hans Albert and grandson Bernhard Caesar on the steps of the garden terrace, 1932

Fig. 1.15. Einstein and the Indian poet Rabindranath Tagore in Caputh, 1930

Despite such famous guests, life in Caputh remained relatively modest. There was no telephone in the house and no radio; no one in the house owned a car. To get there, one first had to take the met train to Potsdam and from there, the bus. On special occasions Einstein would consent to being picked up and delivered home again by one of his wealthy car-owning friends or admirers. The Einsteins were on friendly terms with their neighbors and a few of the villagers, buying fruit and vegetables over the garden fence, in emergencies also sometimes stopping by to borrow their phone. Nevertheless they remained relatively aloof from the concerns of the village—aside from visiting the local pub or attending a fair or other public festivity in Caputh, activities that were necessary then as now for Berliner holiday-makers to be truly integrated within the village community.

The only luxury that Einstein allowed himself was sailing. Wealthy friends and admirers had given him a light cruiser for his fiftieth birthday. He used the seven-meter-long sailboat with a sail surface area of about twenty square meters very frequently in the summer of 1929, calling it affectionately "my chubby sailboat" and naming it *Tümmler* (*Porpoise*). While the house was still under construction during the summer 1929 and the Einsteins were living in the village

Fig. 1.16. Albert Einstein and Adolf Harm, the builder of the boat *Porpoise*

on Potsdamer Strasse, Elsa wrote her sister-in-law Maja in Switzerland: "Our boat is magnificent; Albert has his own landing from the garden, he's enjoying this sailing bliss thoroughly. The boat was a gift of very rich friends (15,000 marks). I write you this so braggingly so that you have an idea what a fine boat your brother sails" (22, p. 304).

The boat's mooring at the "Schumann Wharf" on Potsdamer Strasse was at that time also easily accessible by a path leading directly from Einstein's summerhouse.

But Einstein was able to enjoy navigating the Havel lakes for only four summers and his house, only three. When it was shuttered up for the winter at the beginning of December 1932 and the Einsteins returned to their city apartment briefly before their departure for a lecture tour in Pasadena, California, it was the last time they were to see it. Einstein would not and could not return to a Germany dominated by the Nazis. In the following summer the house and the sailboat were illegally confiscated. The house was first used by a neighboring country home for Jewish children; later it became a kindergarten facility and an army hostel for the German Army. After 1945 it served for decades as a residence for the community of Caputh. Postwar shortages and a lack of interest by the East German state authorities led to the deterioration of this prominent piece of real estate. By the mid-1970s it was in a sorry, dilapidated state. The elaborate celebrations on the centennial of Albert Einstein's birth in 1979, organized by the Academy of Sciences and the East German government, rescued it from threatening collapse. It was placed under the authority of the administrators of the Academy of Sciences for renovation of the building and for making it available for academic purposes. Until the reunification of the two Germanies in 1990, it was used as a guesthouse and conference venue for prominent scientists. With reunification came a reorganization of the East German Academy and the return of confiscated property to the original owners. This new development applied to Einstein's summerhouse, busying lawyers and the courts for years. Finally, the claims of Einstein's heirs were acknowledged, the most important of whom is the Hebrew University of Jerusalem. It designated the Einstein Forum in Potsdam as its agent and manager of the property. A general reconstruction of the building was done in preparation for the Einstein Year 2005, and it has been operating since May 2005 as a meeting center that is also open to the general public.

CHAPTER TWO

Einstein's Workplaces in Berlin

"Bright people attract bright people." This maxim applies today no less than it did yesterday. But the brains of science of a century ago conducted their research not at MIT in Cambridge or at Stanford in Silicon Valley but in Berlin. During the decades at the turn of the twentieth century, the city could boast an abundance of world-famous scholars and excellent research institutions, particularly in physics. The overall development of physics of that period became intimately connected with the research being conducted in Berlin. So it is surely legitimate to consider it, if not an absolutely necessary consequence of this circumstance, at least a consistent one that Einstein would be enticed to come to Berlin by the leaders in his field as the rising star among them in 1913. As a full-time member of the academy in Berlin, Einstein was able to devote all his effort to the problems of interest to him as a so-called paid genius. But his professional activity was not limited to this institution. He had connections with many other scientific establishments of the city as well, ranging from the university at which he lectured to the Kaiser Wilhelm Society (Kaiser-Wilhelm-Gesellschaft), whose Institute of Physics he directed, to the Imperial Institute of Physics and Technology (Physikalisch-Technische Reichsanstalt, or PTR, Germany's bureau of standards), or the research laboratories of the major German electrical combine Allgemeine Elektrizitäts-Gesellschaft (AEG), where he performed experiments as a visiting scientist. In addition, he was involved with a number of other institutions in and around the city in his capacity as a science administrator—whether as president or member of the board of the German Physical Society (Deutsche Physikalische Gesellschaft, or DPG), as trustee of the Einstein Tower and the Astrophysical Observatory in Potsdam, or as science popularizer at today's Archenhold Observatory in Berlin-Treptow.

Site 5. Prussian Academy of Sciences
Unter den Linden 8
10117 Berlin (city center)

subway stop Hausvogteiplatz (U2) or
subway/met. train sta. Friedrichstrasse
(U6, various S lines)
from either, 5 min. by foot

"At Easter I'm going to Berlin as an Academy person without any duties, somewhat like a live mummy. I'm looking forward to this difficult profession" (1, vol. 5, p. 538), Albert Einstein wrote this to his friend Jakob Laub in summer 1913. That same month Einstein was elected a member of the Preussische Akademie der Wissenschaften (Prussian Academy). With the kaiser's sanction in November 1913, his election became official.

Einstein formally accepted the appointment on December 7, 1913, and informed the academy, "I wish to take up my duties during the first days of April." At the same time he thanked his Berlin colleagues

> for offering me a post in your midst in which I can devote myself to scientific work free of any professional obligations. When I reflect upon the fact that each working day demonstrates to me the weakness of my thinking, then I can only accept the high distinction intended for me with a certain trepidation. But what encouraged me to accept the election was the thought that all that can be expected of a person is that he devote himself with all his might to a good cause; and I do feel capable of that. (21, vol. 1, p. 101; 1, vol. 5, trans. pp. 169)

The Prussian Academy, founded by Gottfried Wilhelm Leibniz in 1700, was the largest and most important among the German academies and a typical learned society. The focus of its work lay on discussions and assessments of research findings made by its members, usually at other establishments not directly connected with the academy. It was at these other research establishments in the city, such as universities, that these members worked and earned their salaries as professors or directors. The academy merely paid its members a token "honorary wage" of 900 marks annually. The some seventy academy members, subdivided into two classes, one for the physical and mathematical sciences and one for philosophy and history, met regularly, normally every fortnight, to discuss the latest results in research and formulate statements about issues of science and research policy. The academy was thus a place not just where new findings were presented and discussed but also where strategic ideas for the design of the scientific enterprise were debated. Although the academy's prestige and effectiveness were somewhat limited by a lack of research facilities of its own, election as an academy member was nevertheless considered a high professional and social distinction both nationally and internationally.

It was an even higher distinction to be a full-time staff member, as Einstein had been become. Such academy employees were free to devote themselves entirely to their own projects, without being laden with other obligations. The academy had only two such privileged lifetime positions, one for each of the classes, and they were offered to the most renowned representatives in their field. The full-time academic position for the physical and mathematical class was held by the Dutch physical chemist Jacobus Henricus van't Hoff until his death in 1911. Then it was first offered to Wilhelm Carl Röntgen. But these initial plans fell through because Röntgen preferred to remain in Munich. At the instigation of Walther Nernst and Max Planck, the position was then offered to Einstein, but this decision was preceded by lengthy negotiations because there were questions about the funding. To make it attractive to Einstein not just intellectually but also financially, the compensation was set at the maximum pay level of a German university professor. Einstein's annual salary of 12,000 marks was certainly decent, particularly for a young professor at just thirty-five years of age. But he was certainly not among the highest earners. Other professors made considerable sums from the attendance and examination fees charged to their students, sometimes bringing in an annual income of over 20,000 marks. The base salary of Einstein's colleague Walther Nernst, for instance, at that time came to over 15,000 marks, not including his teaching fees and other professorial supplemental income.

A few days before Einstein settled in Berlin, on March 22, 1914, the academy had taken its rightful place on the thoroughfare Unter den Linden. For over a decade, since 1903, it had been located at Potsdamer Strasse no. 120 while new facilities were being built. The Prussian "court architect" Friedrich von Ihne had been commissioned with the design and construction of the monumental complex for the Royal Library, now called the Staatsbibliothek Unter den Linden. At that time it was the largest library structure in the world. The Prussian Academy of Sciences moved into fine rooms in the wing facing the street. Office space for administrators and the presidents of the academy also accommodated its various commissions and long-term projects. Most impressive of all was the hall for the plenary meetings and the meeting rooms for the two classes. No

Fig. 2.1. Building complex of the State Library, on the thoroughfare Unter den Linden

room was made available to the new full-time staff scientist; whether this was because of a shortage of space or at Einstein's own bidding is not known.

Thus Einstein's place of work in Berlin was also his place of residence. He made his way to Unter den Linden only for the meetings of the academy, its committees, or other boards. As can be gathered from the ironic report of Karl Willstätter, the director of the Kaiser Wilhelm Institute of Chemistry, more attention was paid to academic or social matters than to the scientific agenda at these fortnightly academy meetings:

> At the entrance one equipped oneself with the just recently issued evening paper; the tables offered every convenience for correspondence, in the niches sofas beckoned for one-on-one discussions. The academy was the meeting place for any consultation and conversation, indispensable considering such large distances as for us Dahlemers. It was generally not common to give the talks much attention.

Some speakers turned their backs on the audience, murmuring as they scribbled away on the blackboards. (94, p. 231)

The reminiscence by the sinologist Otto Franke, professor at the University of Berlin, gives a slightly more differentiated picture: "Whoever attended the public meetings of the academy, during which account was made of the ongoing and planned projects got a notion of what the academy did and how worthwhile she was. But unfortunately very little use was made of this opportunity and its offended opponents avoided it on principle" (81, p. 160).

Although Einstein surely did not find the academic pantheon he was hoping for, he took this milieu lightly, which for him undoubtedly called for some adjustment. In a letter to his student Otto Stern he characterized the academy as "amusing, actually more droll than grave. This type of thing is always subject to mass psychology" (1, vol. 8, trans. p. 23).

To his former teacher and colleague at Zurich, Adolf Hurwitz, he reported in May 1914: "In its habitude the academy entirely resembles any faculty. It seems that most of the members restrict themselves to displaying a certain peacocklike grandeur *in writing*, otherwise they are quite human" (1, vol. 8, trans. p. 13).

Years later, in a letter to Heinrich Zangger, even Einstein himself became the target of his own keen ridicule: "The intellect gets crippled and one's strength dwindles away, but glittering renown is still draped around the calcified shell. . . . I am just right for the academy, whose quintessence lies more in sheer existence than in activity" (1, vol. 8, trans. p. 622).

Such ironic self-stylizations should not disguise the fact that Einstein took an active and intense interest in the academy's affairs. He not only made regular appearances at the academy meetings but was a disciplined participant at its committee meetings. Most important, the academy became Einstein's place of choice for public announcements to his fellow scientists about his latest results. Einstein's academy publications, talks, and protocoled discussion remarks number more than fifty. That means that every year Einstein presented between two and five scientific papers at the meetings of the academy or submitted them for printing in their proceedings. The height of his activity there was unquestionably the fall of 1915, when after many years of struggle, he was able to complete his general theory of relativity. At the meeting of November 4, he presented the new theory before the academy (and added a few alterations to it the following week). On November 18 he informed the academy during its plenary session about a paper in which the perihelion motion of Mercury's orbit is explained on the basis of his revised theory—one of the three new consequences of his theory

Fig. 2.2. Meeting of the academy; in the front row, second from the left is Albert Einstein

and otherwise not explicable by the classical Newtonian theory of gravitation. Finally, on November 25 he published the corrected field equations of gravitation. He wrote to his friend Zangger the next day, full of pride: "The theory is beautiful beyond comparison" (1, vol. 8, trans. p. 151).

Enthusiasm about Einstein's accomplishment was muted among his Berlin colleagues, however. Einstein's efforts to generalize his theory of relativity was long regarded by his fellow physicists with great skepticism, if not downright opposition. Max Planck, otherwise an enthusiastic supporter of Einstein, formulated this in his laudatio on the occasion of Einstein's election to the academy with characteristic restraint: "Currently he [Einstein] is working intensely on a new theory of gravitation; only the future can instruct us with what success" (64, vol. 1, p. 96).

A year later, in reply to Einstein's inaugural speech, Planck became a little more explicit and spoke of the threat of Einstein "occasionally wandering too far off into obscure areas" (64, vol. 2, p. 247).

Unimpressed by such skepticism, Einstein held his course, and it must have been a private triumph for him to be able to stand before his doubters again just one year later and set them straight. He reflected on this, just after his election to the academy, in a letter to his friend Michele Besso with his typical irony:

"The fraternity of physicists behaves rather passively with respect to my gravitation paper.... Laue is not open to the fundamental considerations, and neither is Planck, while Sommerfeld is more likely to be so. A free, unprejudiced look is not at all characteristic of (adult) Germans (blinders!)" (1, vol. 5, trans. p. 374).

Only when the solar eclipse expedition led by the British astronomer Arthur S. Eddington succeeded in verifying another effect of the theory, gravitational light deflection, did the tide turn in the reception of Einstein's theory. Einstein not only gained acceptance among his colleagues but suddenly became a public figure of world renown. By then it was very clear that Einstein's work on the general theory of relativity was one of the highlights in the history of the academy and, indeed, of Berlin science in general.

Einstein was a challenge for his fellow academicians not only in scientific matters but also in politics. His attitude sharply contrasted with that of the overwhelming majority of German physicists. When World War I broke out, he did not join their hurrahs. Einstein regarded the patriotism and chauvinism prevalent among academics as "a kind of mental epidemic": "Only very uncommon, independent characters can escape the pressure of prevailing opinion. There seems to be no such person at the academy" (1, vol. 8, trans. p. 314).

Friedrich Meinecke remembered a conversation he had with Einstein during the war as they were walking down the street Unter den Linden: "His despair about the war and his extreme, pacifist mentality became so glaringly obvious that I, critically and antinationalistically inclined though I myself then was, felt noticeably estranged him" (89, p. 185).

When the revolution of November 1918 took place, Wilhelm II was forced to abdicate, and the first German Republic was declared, Einstein had great hopes about social and political improvements. Once again he was at odds with his colleagues: "Among the academicians, I am some kind of high-placed Red [*Obersozi*]" (1, vol. 8, trans. p. 944), he wrote in November 1918 to his mother.

As a prominent pacifist, democrat, and Jew, Einstein's scientific work took on a symbolic political significance. His relativity theory was seen as the incarnation of "Jewish" and "Bolshevistic" science and became the object of chauvinistic and anti-Semitic slander campaigns. Even when some of these attacks threatened Einstein personally in the 1920s, the academy did not deem it necessary to speak out publically in support of him or to repulse the aggressive assaults. Such silence was symptomatic and consistent with the academy's self-image founded on a separation of science and politics.

The conduct of the academy and its representatives in 1933 was more mortifying than brilliant. When power was conferred to the National Socialists on January 30, 1933, Einstein happened to be away on a research stay in California, but he followed the news about the events in Germany with great attention. When he became certain that the policy of the new ruling party included brutal force and political infringements and that his fellow Jewish "clansmen" and other political opponents of the Nazis were being subjected to merciless harassment, discrimination, and persecution, he did not hesitate to speak out. At the beginning of March—just before his departure from Pasadena—he made the following unmistakable declaration: "As long as the possibility remains open to me, I will only stay in a country in which the political freedom, tolerance, and equality of all its citizens prevail before the law. . . . These conditions are currently not satisfied in Germany" (5, p. 81).

With this public statement against Nazi Germany and avowal of basic democratic rights, Einstein became one of the few prominent German scientists to raise their voices immediately and uncompromisingly against the Nazi terror and to protest against the injuries to the civil rights of the Jewish minority and other disfavored countrymen.

Einstein's statement predictably raised indignant protests by representatives of the Nazi regime who cared about the new government's international reputation. They had already cast insults and mean-spirited attacks against Einstein, now a target of preference. Most of his colleagues at the academy could not understand how he could choose such unequivocal words, even discrediting them as participation in a "foreign slander campaign" against Germany. Einstein thus became "awkward," not just for Nazi Germany but also for the academy. His academy membership was to be used as a political signal, and plans were in place to demonstratively throw him out. But Einstein anticipated their action and informed the academy of his resignation in a letter dated March 28, 1933. Max von Laue recalled how furious the Ministry of Culture was "at having forestalled them with his resignation" (21, vol. 1, p. 273).

Although the line separating Einstein from his fellow academicians had been clearly drawn by his decision, it was certainly not easy for him to turn his back on Berlin forever, deeply disappointed though he was by their opportunism. "For 19 years," he recalled in his resignation letter to the academy,

> the academy has given me the opportunity to devote my time to scientific research, free from all professional obligations. I know how very much I am obliged to her. I withdraw reluctantly from this circle, also because of the intellectual stimulation and the fine human relationships which I have enjoyed throughout this long period as a member and have always valued highly. (21, vol. 1, p. 246)

The response of the plenum of the academy to Einstein's resignation was the terse statement that "with the resignation of Mr. Einstein, further measures by her are obviated" (21, vol. 1, p. 246). In a press release on April 1, not coincidentally the day that was dubbed "national Jew boycott day," the officiating secretary Ernst Heymann even added that the academy had "no cause to regret Einstein's resignation" (21, vol. 1, p. 248).

A few days later Max von Laue tried to play down this embarrassment and dampen the academy's ignoble conduct somewhat by acknowledging how much the academy had lost with the departure of a member of such immense scientific importance as Einstein. Still, the eagerness with which his academy colleagues fell into line was one of the most painful things Einstein ever experienced. All too willingly did they bow to the political pressure and allow their tradition-rich institution to be manipulated according to National Socialist precepts. Einstein's disappointment is evident in a letter written that April to his erstwhile supporter and fatherly friend Max Planck, whose reactions at the time had also left much to be desired, as Einstein reminded him in an outburst of bitterness "that, all these years long, I was just useful to Germany's reputation; and that I never gave a thought to the fact that, with the systematic agitation going on against me in the rightist press—especially in recent years—no one thought it worth their while to stand up for me" (8, p. 233).

For Einstein this resignation severed his bond with the academy—indeed, with Germany—once and for all. After the fall of the Third Reich, with all the atrocities committed during that period in the name of Germany, Einstein wanted nothing more to do with this "land of mass-murderers." He consistently declined all invitations by his former colleagues and requests to resume former memberships, not always choosing a friendly tone. In the summer of 1946 he dismissed the attempt by Johannes Stroux, the first postwar president of the Prussian Academy, to convince him to rejoin, saying, "After all the horrific things that have happened, I consider myself incapable of accepting the kind offer by the German academy" (23, p. 527).

In the same period he wrote quite bluntly to his Munich colleague Arnold Sommerfeld in reply to a similar request by the Bavarian academy: "After the Germans have slaughtered my Jewish brothers in Europe, I want nothing more to do with Germans, not even with a relatively harmless academy" (15, p. 121).

1. Your fine and tender little letter
   rang so true to German manner;
   Wanting not to stay in debt,
   down I wrote these verselets.
2. In deep dismay did I gather
   that I did you all forestall,
   so the careful court of justice
   could not reach me with its gall.
3. He who tells a grisly story
   will be judged by us in wrath.
   Tho' it be the truth he says,
   impossible, then, if we forgave.
4. So much fun it is for us,
   to tell a fib in corpore.
   Ficker sets it in good form.
   Yes! That'll hit him like a storm!
5. German honor knows he not,
   Asia's damned good-for-naught!
   Prussian highness gone to waste
   on that headstrong willfulness.
6. On him is wholly lost, alas,
   the fervor of a patriot,
   although he had the benefit
   of our lengthy eloquence.
7. Bold we are, now and then,
   from the safety of our den.
   But before the rabble's cudgel,
   our wobbly knees may tend to buckle.

1. Euer Briefchen fein und zart
   Klang so traut nach deutscher Art;
   Weil ich nichts wollt' schuldig bleiben,
   Tät ich diese Verschen schreiben.

2. Tief betrübt hab' ich vernommen,
   Dass ich Euch zuvorgekommen,
   Dass ich konnte treffen nicht
   Wohlerwog'nes Strafgericht.

3. Wer da Greuelmärchen dichtet,
   Grimmig wird von uns gerichtet.
   Wenner gar die Wahrheit spricht,
   Dann verzeihen wir's ihm nicht.

4. Freilich macht es viel Vergnügen
   Uns, in corpore zu lügen.
   Ficker bringt's in gute Form.
   Ha! so trifft man ihn enorm!

5. Deutsche Würde kennt er nicht,
   Asiens verdammter Wicht!
   Preussens Hoheit, strammer Geist
   sich verschwendet da erweist.

6. Von der Vaterländerei
   Blieb er leider gänzlich frei,
   Trotzdem lange unverdrossen
   Unsere Reden er genossen.

7. Mutig sind wir dann und wann,
   Wenn uns nichts passieren kann.
   Doch vor mächt'gen Pöbels Knute,
   Wird uns manchmal schwach zumute.

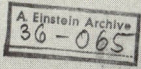

Fig. 2.3. Albert Einstein's scornful rhyme in honor of his fellow academicians from 1933

Site 6. Humboldt University
Unter den Linden 6
10117 Berlin (center)

subway stop Hausvogteiplatz (U2) or
subway/met. train sta. Friedrichstrasse
    (U6, various S lines)
from either stop, 5 min. by foot

As a member of the Prussian Academy of Sciences, Albert Einstein also had the right to offer lectures at the University of Berlin, which at that time was named after the Prussian king Friedrich Wilhelm III, who had founded the university in 1810, as Friedrich-Wilhelms-Universität. It was a right but not a duty, with the purpose of strengthening the bond between the academy and the university. As a full-time academy scientist, Einstein found this right particularly valuable because it secured him the status of a university professor, so prestigious and influential as that title was, particularly in Germany. In this capacity he was free to announce any courses he wished to teach and was also a fully authorized member of the faculty; that meant he could conduct examinations, advise doctoral candidates, and have an equal say on committees of the university. But Einstein's administrative involvement as a faculty member remained mainly restricted to participation in a few appointment proceedings. For example, he exerted his influence on the professorship appointments of the philosopher-scientist Hans Reichenbach and the founder of wave mechanics Erwin Schrödinger at the University of Berlin.

Einstein started teaching at the university right after he had settled in Berlin. He announced a lecture on the "theory of relativity" for the summer term of 1915, as he would repeatedly do in the coming years. Other topics included "statistical mechanics and Boltzmann's principle," "statistical mechanics and quantum theory," and "various topics in theoretical physics." This was evidently not a particularly abundant or original program of lectures, but it was designed primarily for a specialized group of students and was only offered during the first half of his Berlin period. From the mid-twenties onward, the only course announcement we find is a "physics proseminar" that he offered (if at all) together with his colleagues Max von Laue, Wilhelm Westphal, and Gerhard Hettner for higher-level advanced students. The term "proseminar" was then used to describe courses ranging from group sessions to private consultations between students and their supervisors focused on preparing their doctoral theses. As no students are known to have taken a degree under Einstein's guidance at Berlin, his teaching load in this regard was clearly very limited. After 1930 the university's course catalog no longer lists any offerings by Einstein. This is explained by the fact that Einstein had by then started to teach lecture cycles over the course of a few months every year in Pasadena, California.

The reports and opinions about Einstein's classes and teaching skills strongly diverge. Robert Rompe remembered Einstein's lectures on relativity theory to an audience of less than ten individuals, and

> of these 1 or 2 were teaching assistants. It was naturally very impressive how he spoke and how he set the accents. But perhaps it was too challenging for us at our youthful age. He usually broke the lecture off at a place (with a reference to the literature, above all to the pertinent book by Max von Laue), where the matter would have demanded longer derivations and calculations. (72a, p. 47)

Philipp Frank, Einstein's successor at Prague, pointed out in this regard that his lectures were

> not easily accessible even for students whose specialty was physics. For the majority of students everywhere demand that one spoon-feeds them with precooked

Fig. 2.4. The Friedrich Wilhelm University with the *Kommode* (the Royal Library until 1914), around 1930

food, whereupon they often do not even take the trouble to digest it themselves. For such auditors Einstein's lecture course was—indigestible. And auditors who wanted to think these difficult problems through themselves, and were able to do so, were never very many, not even in such large cities as Berlin. (26, p. 332)

Nevertheless, there were times when Einstein's lectures and talks drew such crowds that they had to take place in the main auditorium of the university. This was because at the beginning of the 1920s Einstein had become a public figure, one might say science's first "pop star." A true "Einstein rumpus" prevailed, and every self-respecting paper reported about the spectacular theory and its creator. This unleashed an unprecedented invasion on Einstein's courses in the early 1920s, and contemporaries recalled: "in those days one did not even have to ask in which room Einstein was lecturing, just to follow the stream of people going to it at the given time" (26, p. 331).

Even the rabid reporter Egon Erwin Kisch thought Einstein's teaching courses worth an article:

It was almost entirely foreigners who sought out the university, between a visit to the Potsdam palace and the performance at the Metropol Theater, in order to have a look at the "modern Newton." Einstein knew, though, that his auditorium was not full of people with a thirst for knowledge but of the curious, and so he

Fig. 2.5. A posting for the lecture "Theoretical and Experimental Issues on the Question of the Formation of Light," 1927

thought up a good procedure in order not to let himself be the object of blank, uncomprehending gazes for two hours long, without having to turn those away who were in fact interested in the content of his lecture. He opened the lecture with a brief summary of what he had discussed last Tuesday and said: "I am now going to read about this or that law, but will make a little break so that everyone who is not interested in it can leave the hall." When ten minutes later he spoke about this or that law, of the three hundred auditors usually scarcely fifteen had remained behind. (86, p. 329)

As Einstein's popularity grew, so did personal attacks on him and public objections to his theory. The first disruptions to his courses happened during the winter term of 1919–20, when he was lecturing on "relativity theory." The official story was that the student association had protested against auditors of his courses who had not paid the attendance fee. But the anti-Semitic undertone could not be missed. Similar things were to happen again later—such as in the summer term of 1922, when on the heels of Walther Rathenau's assassination Einstein also became the target of more virulent anti-Semitic aggression. When he began to receive death threats, he decided to cancel his lecture. But he promised his students that he would repeat the course in the future. His letter continues:

> I had to interrupt the lecture in the summer term because I had been urgently advised, in the interest of personal safety, not to perform any public functions at this time. In the coming semester I unfortunately cannot lecture because I shall be away from Berlin [on a trip to Japan]. I was pleased that by your request you are showing so much of an interest in the subject; I will strive to recompense what has remained outstanding as soon and as completely as possible. (58, p. 21)

In the following spring Einstein fulfilled his promise with eight two-hour lectures on the theory of relativity, for which more than sixty interested persons had registered. After that Einstein is found listed only twice more in the university's course catalog, for individually arranged sessions on "various topics in theoretical physics" and the aforementioned "Physics proseminar."

Yet Einstein had also been offering a series of popular science talks at the university besides his regular courses. They had been organized by the workers association, a mathematical-physical study group (Mathematisch-Physikalische Arbeitsgemeinschaft, or Mapha), and they usually appealed to his audience much better than his official courses did. It reportedly drew audiences of over a thousand. They covered topics like "geometry and experience" (February

23, 1921), "on the present state of the problem of the nature of light" (February 24, 1922), "the essence and the present state of relativity theory" (January 25, 1926), and "theoretical and experimental issues on the question of the formation of light" (February 23, 1927). Published versions of some of them appeared as well.

Since 1965, the fiftieth anniversary of the general theory of relativity, a plaque has hung in the lobby of the cinema hall (*Kinosaal*) at Humboldt University (entrance off Dorotheenstrasse) in memory of Einstein's lectures and public talks at the University of Berlin.

Site 7: Physics Institute of the
   Friedrich Wilhelm University
Now the ARD national broadcasting
   studio at the capital
Am Reichstagufer 7/8
10117 Berlin (center)

subway/met. train sta. Friedrichstrasse
   (U6 and various S lines)
or met. train sta. Unter den Linden
   (S2, S25)
from either stop, 3 min. by foot

Between 1873 and 1878 a complex of scientific institutes was built for the university on the site of the abandoned artillery workshops on the riverbank close to the Brandenburg Gate. The Physics Institute was located on the side near the River Spree. This institute was for a long time among the largest and most important in its field in Germany. The list of institute directors, including names like Hermann von Helmholtz, Emil Warburg, and Walther Nernst, testifies to its excellence. Numerous assistants and other members of the staff also contributed significantly to the institute's reputation: Heinrich Hertz, Otto Lummer, Peter Pringsheim, Erich Regener, Wilhelm Wien, and other important physicists of the nineteenth and twentieth centuries started their scientific careers here. Here also, Eugen Goldstein performed his fundamental experiments on cathode rays and canal rays as a guest researcher, and foreign guests, such as the Russian Peter Lebedev, the American Michael Pupin, and Albert Abraham Michelson, sought to complete their scientific training here. James Franck and Gustav Hertz, then still a private lecturer and a teaching assistant, respectively,

Fig. 2.6. Physics Institute on the riverside road, Am Reichstagufer

began conducting their gas-discharge experiments at the institute in 1911, which ultimately earned them a Nobel Prize. Indeed, one could say that the former institute complex at Am Reichstagufer is the spot with the highest "Nobel Prize density" in Berlin.

Albert Einstein also had business at this location, but not as a researcher. He did not require the institute's excellent facilities for his theoretical research, and his employers were the academy and the Kaiser Wilhelm Society. What drew him regularly to this building on Am Reichstagufer was the fact that its large auditorium was the place where the Wednesday colloquium, so famous particularly in the 1920s, met. At that time it was supervised by Max von Laue and took place every Wednesday between 5:00 and 7:00 p.m. Any self-respecting physicist in the city made sure to be among the audience every week. The first row of the auditorium was often filled with Nobel laureates. An invitation to speak before this audience was consequently a high honor, and it was a good opportunity to find out about current advances and problems in physics firsthand.

Einstein was not just a regular member of the audience. He also appreciated the discussions afterward, considering them a unique opportunity. After attending the colloquium, he supposedly repeatedly told his wife, Elsa, "One probably doesn't find such a cluster of excellent physicists anywhere else in the world today" (26, p. 387). Einstein influenced the colloquium directly himself as a speaker, but even more so through the originality of his contributions during the discussions that followed. On the occasion of the Einstein centennial celebration in 1979, Robert Rompe remembered:

> At Max von Laue's colloquium Einstein played a prominent role. Not that he gave long and broad-ranging commentary on the speaker's presentation . . . , rather because he had a short question for almost every talk, not with the goal of correcting the speaker or teaching him a lesson, rather simply to express his interest in the talk and the subject of the talk. His mere presence at the colloquium raised the level, so to speak; because he took part, it was interesting. (72a, p. 47)

The Physics Institute was not only the site of this colloquium whose influence reached far beyond the city's limits but also the seat of the German Physical Society. Since this society had far fewer members during Einstein's time than it does today—in the 1920s its membership numbered not much over a thousand—a special building for it was not yet necessary. One business office where the meetings of the board and other consultations could be conducted was sufficient. At least between 1914 and 1925 Einstein's presence would have been frequent in his capacity as a member of the society's board. A few weeks after his arrival in Berlin, at the meeting of May 8, 1914, he was elected to sit on the society's board. This is another indication of the high hopes placed on Einstein's appointment in Berlin for science and its politics. Although the society had given itself the name Deutsche Physikalische Gesellschaft in 1899, it was dominated by its mother organization, the Physikalische Gesellschaft zu Berlin, until well into the 1920s. Members of the latter comprised the majority of the membership of the former and until that time also provided all the DPG's presidents, which was cause for resentment by the non-Berliners, particularly a splinter group of Bavarian physicists headed by Wilhelm Wien. An attempt was made within the DPG itself to quell these bad feelings and prevent a threatening split by proposing new statutes for the society. During the consultations about the new statutes Einstein was apparently involved in a special way as a freshly elected member of the board and as a "neo-Berliner." Since, as he wrote Wilhelm Wien, he "cannot have become a 'Berliner' yet in such a short time" (1, vol. 8, trans. p. 24), he tried to act as an objective mediator between Berliner and non-Berliner interests. With these efforts Einstein evidently succeeded in gaining the respect and recognition of both sides. Not only was his seat on the board confirmed the following year, but he was even elected to succeed Fritz Haber as president of the society at the meeting of May 5, 1916. The fact that many physicists were then "on the battlefield" or were performing some other service related to the war effort, from which Einstein as a Swiss citizen was exempt, must have played a role in his election. He held this office as president of the society for the duration of its two-year term, until 1918.

In his new function Einstein had to act as the representative of the society toward the outside world and direct its operations. In the latter he was ably supported by Karl Scheel, physicist at the Imperial Institute of Physics and Technology (PTR), who served as secretary or managing director of the society. Even though the society did not have as much such business to settle at that time as scientific societies now have, the presidency still required quite a lot of commitment and organizational skill. Specifically, it was the president's responsibility to chair the meetings of the society, which convened every other Friday. These meetings dealt with the ordinary business of the society, that is, addressing specific problems or honoring recently deceased members, and it was the president's task to compose a brief commemorative text. But these meetings were also as much occasion for scientific debate and presentations as were the Wednesday colloquium. For example, Max Planck presented his radiation law and his revolutionary quantum hypothesis at meetings of the DPG in the fall of 1900.

Einstein himself delivered more than twenty such talks over the course of his almost two-decade membership (he joined the society in November 1913, before he came to live in Berlin). Their scope spans Einstein's entire oeuvre, ranging from quanta and relativity theory to thermodynamics and mechanics. Experimental issues were also treated, particularly concerning the discovery of what is known as the "Einstein–de Haas effect." Einstein had achieved this with Wander Johannes de Haas soon after the Dutch physicist had come to work in Berlin as a guest researcher at the PTR. Einstein devoted four talks to this topic alone in 1915, and he also reported on the subject at the Physical Society.

Another highlight of the meetings of the DPG during that period was the colloquium in celebration of Max Planck's sixtieth birthday. As president of the society, Einstein was responsible for organizing it. The speakers at this festive meeting on April 26, 1918—who included, besides Einstein himself, his fellow physicists Max von Laue, Arnold Sommerfeld, and Emil Warburg—enthusiastically applauded Planck's lifework, acknowledging its central importance in the development of physics in Germany. These Planck festivities were the crowning event of Einstein's presidency, as can be gathered from the published reports as well as from Marga Planck's letter to Einstein thanking him on behalf of her husband for the "fine evening" and for the copy of Einstein's "sermon," his speech on "Motives for research": "We delighted in your thoughts. . . . Also—this

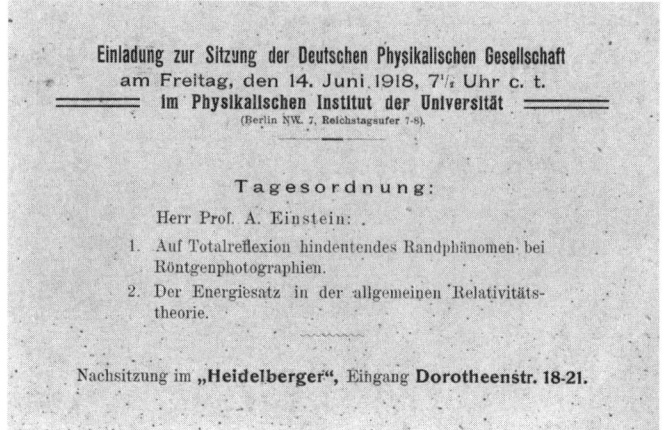

Fig. 2.7. Advertisement for one of Einstein's talks, 1918

I would like to express to you today—I personally am quite especially pleased that my husband has found such a warm friend in you!" (1, vol. 8, trans. p. 545).

Although as a former president Einstein remained a member of the society's board until 1925, he gradually reduced his involvement in the society's affairs at the beginning of that decade. He had already begun to feel estranged from his role in Germany and Berlin in particular. He frequently absented himself, going abroad on extended travels. Allowing himself to be harnessed as a scientific representative and organizer was at odds with his personality—regardless in whose service.

No longer on the board but still a prominent regular member of the society, Einstein joined the initiative in support of establishing a medal award. He added his signature to the related appeal at the end of 1927. The foundation was dedicated to Max Planck in honor of his seventieth birthday, and on the golden anniversary of his doctoral degree, on June 29, 1929, the first award ceremony took place at a festive meeting of the society in the Physics Institute. On this occasion the first Planck Medal was conferred on its namesake by the officiating president of the society, Hermann Konen, and then Einstein received the second medal from Planck's own hands. Einstein thanked him for the honor, which to this day is the highest distinction of the Physical Society, with a very personal and moving speech.

A very subjective report about this day was written by Einstein's friend and doctor, János Plesch:

Einstein had eaten lunch with me, had, after a black coffee, lain himself down and slept soundly. The meeting was at five o'clock in the afternoon. At about half past three Einstein said: "I'm probably going to have to say something to my people this afternoon," . . . [sat] himself down at my desk, and [requested] a piece of paper. I let him use whatever was lying on the desk and he happened to pick up a bill by my shoemaker. He filled the blank back of it up with his writing in about twenty minutes. We then went to the Physics Institute where the festive session was to take place. The hall was occupied to the last seat with famous personalities of the mathematical sciences. Planck emphasized how proud it made him to be able to award the medal named after himself to a scientist of such merit, etc., etc. Then Einstein spoke and said: "When I accept such honors I always become emotional and so I wrote down what I want to say to you as thanks—I shall read it to you." He

Fig 2.8. Max Planck and Albert Einstein, the first recipients of the Max Planck Medal in 1929

then put his hand in his pocket, pulled out the shoemaker's bill, and read about the crisis of causality.... Thus everything continued on not without a touch of human emotion—yet extremely matter-of-factly ... [;] after the meeting Einstein gave me the medal; it was a lump of gold with Planck in relief. Einstein had looked at the medal only in its case, he had not removed it or taken it in his hand. I had it for years in safekeeping until I gave it back to his wife Elsa. Einstein himself had never asked for it. That evening he went out with Slevogt, Grünberg, and me to a genuine Munich restaurant. Not a word more was lost on the venerable session. (70, p. 115)

As did his affiliation with the Berlin academy, so also did Einstein's membership in the DPG ended in the spring of 1933. It was clear to him that after his spectacular withdrawal from the academy the Nazis would apply the same political pressure on the DPG and other institutions to which he was connected. So he wrote his friend and colleague Max von Laue on June 5, 1933:

> I heard that my unsettled relationship with German corporate bodies in which my name still appears on the membership lists could cause difficulties for friends of mine in Germany. That is why I ask you please to make sure sometime that my name be struck from the lists of these bodies. They include, e.g., the German Physical Society. (96, letter; 47, trans. p. 288)

Site 8. Prussian Ministry of Culture
Ministry of Religious, Educational, and
  Medical Affairs
Wilhelmstrasse 16
10117 Berlin (center)

subway stop Mohrenstrasse (U2)
or met. train sta. Unter den Linden (S2, S25),
from either stop, 3 min. by foot

Einstein's presence in Berlin was not solely the result of initiatives of scientists like Fritz Haber and Max Planck. It was also due to the cultural administration and the willingness of its competent officials to lend an ear to such a plan. Friedrich Althoff, the "omnipotent Ministerialdirektor" of the Prussian Ministry of Culture had a unique modernizing influence on science policy in Germany during the decades straddling the turn of the century. Known as the "Althoff system," this influence endured well beyond his death in 1908. It was based less on a set of theoretical guidelines than on a will to support science and to provide it with an efficient and flexible structure. Althoff was reacting to the demands of a time characterized by industrialization, a steadily growing need for specialists in industry and management, and an increasing professionalization of many fields of work. In this landscape, under the aegis of Althoff's policy, the German, and specifically the Prussian, system of higher education underwent comprehensive reorganization and systematic development, as did scientific activity overall.

In the process the hitherto generally independently operating fields of science, education, politics, management, and business were interconnected and the system of higher education extended and reorganized into an academic mass enterprise. Expansion was not the sole emphasis of this science policy, nor were budgetary hikes or other increases in material resources. The focus was turned on science. This change was connected with occasionally unbureaucratic and strongly autocratic promotion of the highly skilled and talented,

Fig. 2.9. The former Prussian Ministry of Culture, Wilhelmstrasse, at the corner of Behrenstrasse

thus earning Althoff the deferentially ironic nickname of the "Bismarck of academia." Although this style of administration was at odds with democratic ways, Althoff's office had success on its side. He managed to push through the appointments of numerous young and highly skilled scientists—at times even against the vote of the faculties. The list of those who profited by his support includes quite a few future Nobel laureates, such as Robert Koch and Emil von Behring. Max Planck was appointed by Althoff to the University of Berlin in 1889 at the young age of thirty.

After Althoff's death in 1908, his legacy, the "Althoff system," was continued by his closest colleague, Friedrich Schmidt-Ott, and other officials steeped in his tradition at the Ministry of Culture. One of these officials wrote in his memoirs: "In the 1920s, young officials of the Prussian Ministry of Culture still [learned about] Prusso-German university policy by reading through Althoff's files" (78, p. 36). One of these "élèves" of the Althoff system was Hugo Andres Krüss, who had been personally hired by Althoff to serve as a young assistant at the Prussian Ministry of Culture and had closely followed Einstein's academic career since 1912. Krüss was born in Hamburg in the same year as Einstein, 1879. His father was the instrument-maker Hugo Krüss. After studying physics at Jena, where he completed a doctorate in 1903, he was engaged by the preparatory committee for the World Exhibition in St. Louis, where he subsequently directed the "scientific instruments in mechanics and optics" section of the German pavilion. The organizational skills and ability he showed as a science policy maker attracted Althoff's attention, and the twenty-eight-year old was entrusted a position at the Prussian Ministry of Culture. There he rapidly advanced from an assistantship to a professorship in 1909, becoming a privy government councillor and rapporteur in 1918, later to be promoted to department manager in 1920 and department head in 1922. Finally, that year he was appointed director of the university division. In 1925 he was named managing director of the State Library but remained active as a science policy maker—particularly in the area of international relations. His influence grew internationally not only in the library sciences but also from his activities in the League of Nations. In this capacity his contact with Einstein continued, since as the German representative he filled various functions for other international organizations. Sometimes he worked together with Einstein as his substitute or successor, for instance, on the International Committee on Intellectual Cooperation of the League of Nations. But the two men did not always see eye-to-eye; Krüss did not enjoy Einstein's complete confidence. But their differences were not just of a personal nature. They were rooted in fundamentally divergent views on politics and the world in gen-

Fig. 2.10. Hugo Andres Krüss

eral. This became sorely obvious in 1933 when Einstein raised public protest against the onset of political and racist persecution in Germany with the rise of the National Socialists. While Einstein was forced to emigrate as a consequence, Krüss revealed a great willingness to compromise and cooperate with the new regime. In April 1945 he stood literally before the ruins of his lifework in Berlin and chose—like many of his contemporaries—suicide by taking poison.

Whether any regrets about Einstein's expulsion from Berlin played a part in his tragic final reckoning of his life, we cannot now say. When Einstein tendered his resignation at the Berlin academy after being compelled to emigrate in 1933, Krüss showed no sign of sorrow at seeing him leave, let alone protest, despite his direct involvement in Einstein's original appointment to the Berlin academy. He was fully aware of Einstein's importance for Berlin science.

As a member of Schmidt-Ott's staff, Krüss witnessed the founding of the Kaiser Wilhelm Society and from 1911, as government commissioner, attended all the meetings of the society's senate and administrative committee, and he had been involved in helping bring about Einstein's appointment to academy in Berlin. His brief identifying mark, "Kr," is found on numerous official documents of the Prussian Ministry of Culture in matters involving Einstein from the years 1913 to 1925.

The relationship between Einstein and Krüss probably started with Fritz Haber's letter to Krüss in January 1913: "In a conversation about the Ord. Professor of Theoretical Physics at the Polytechnical Institute in Zurich, Dr. Albert Einstein, that we had in the just elapsed year, you brought up the question whether a position could not be created for this extraordinary man at the institute in my charge" (1, vol. 5, trans. p. 327).

As we know, this plan did not materialize, and Einstein did not join the staff at Haber's institute but received a very much more favorable offer of fully paid membership in the Berlin academy as well as directorship of his very own, newly founded Kaiser Wilhelm Institute of Physics. Krüss must also have directly participated in the implementation of this highly ambitious piece of science policy, because his distinctive abbreviation is found on the academy's letter in which the presiding secretaries inform the ministry of Einstein's election (21, vol. 2, p. 17). When the founding of the Kaiser Wilhelm Institute of Physics became a reality in 1917, the working relationship between the two men became closer, as Krüss appears to have been the ministerial official entrusted with the founding affairs and later became a member of the institute's board of trustees, initially standing in for Schmidt-Ott. Krüss was likewise involved in the ministerial aspects of the construction of the Einstein Tower and the Einstein Donation Fund, and in consultation with Einstein he helped bring about the employment of Erwin Finlay-Freundlich at the Astrophysical Institute in Potsdam, subsequently installing him as director of the Einstein Tower (1, vol. 8, trans. pp. 441). Krüss and Einstein also cooperated on other personnel issues—for example, regarding the appointment of a new director for the Geodetical Institute in Potsdam when it was a matter of preventing the selection of a military general for the post. Einstein wrote in this connection, and surely also with Krüss's sanction, to Undersecretary of State Carl Heinrich Becker in the fall of 1919:

> The Geodetic Institute in Potsdam, hitherto one of the most reputable scientific institutes throughout the world in this field, is in danger of coming under the leadership of a general who is but remotely concerned with scientific projects. If this choice materializes, it would inflict serious harm on our scientific interests and even severer harm on the prestige of German science abroad. (1, vol. 9, trans. p. 114)

Einstein and Krüss were successful in their efforts, because instead of the general, the esteemed Ernst Kohlschütter eventually was offered the directorship of the institute in 1922.

All in all, Krüss was clearly the contact person Einstein dealt with when he had business with the ministry. When it was a matter of adjusting the beneficiary information for his widow-pension policy to conform with his imminent divorce from Mileva, there was another intense exchange with Krüss in which not just the minister but Einstein's second wife, Elsa, the person most centrally concerned, was involved.

Krüss proved his diligence and competence as well as diplomatic skill not just in the later years as director of the State Library and as member of various international committees. His talents were already evident at the beginning of the 1920s when he was made departmental head at the Ministry of Culture. His machinations assured that Einstein's Nobel Prize certificate be handed out in Stockholm not to the Swiss envoy but to his German counterpart. Even though he surely knew that Einstein had insisted on keeping his Swiss citizenship when he took the appointment in Berlin, he wrote at the beginning of December 1922—while Einstein was away on his Asian travels—to the Foreign Office:

> In any event, every effort must be made to have the Nobel prize handed over to the German envoy. Professor Einstein is not just German by birth but, above all, as a scientific personality and researcher, a resident of the German Reich—and it is in this capacity that he has received the Nobel prize. He himself repeatedly emphasized that he was a German professor. . . . Based on his own views I do not doubt that he himself would not understand not being treated like a Reichsdeutscher at this occasion so important for the reputation of German science and so visible to the whole world. (47, p. 274)

Einstein's view of the matter was very different, though, resulting in further bickering between the German and Swiss diplomats. When Einstein left for his trip to South America in 1925, Krüss was again the one to try to persuade him to reconsider his evasive attitude toward Germany's diplomatic representatives abroad. At the beginning of 1925 Krüss informed the Foreign Office that "yesterday he had spoken with Professor Einstein about his planned trip to South America and also suggested to him to contact the German envoy, since he was, after all, being paid by Prussia. Prof. Einstein said he was ready to follow the suggestion" (47, p. 282).

Whether Hugo Andres Krüss could thus be regarded as Einstein's advocate at the Prussian Ministry of Culture remains an open question. But he certainly was a capable and competent ministerial official who, true to the Althoff tradition, acted as the "good spirit" of science and was very successful in creating the framing conditions for prominent researchers like Einstein in Germany.

But that alters nothing about Krüss's sometimes contradictory attitudes, which Einstein surely had in mind when he alluded to the "silly and conceited Mr. Krüss" in 1931 (21, p. 351).

**Site 9. Kaiser Wilhelm Institute of Physical Chemistry
and Electrochemistry**
Currently the Fritz Haber Institute
Faradayweg 4–6
14195 Berlin (Dahlem)

subway stop Thielplatz (U1) or
met. train sta. Lichterfelde West,
from there 5 or 10 min., respectively, by foot

The initial plan to attract Einstein to Berlin involved neither the university nor the academy. Neither was there any vacant professorship for theoretical physics or any paid research position available. The plan pursued by the Berlin physicists and ministerial bureaucrats in the winter of 1912–13 was to add this star in the skies of physics to their newly founded scientific institution, the Kaiser Wilhelm Society, as another one of their promising luminaries.

The Kaiser Wilhelm Society, founded in 1911, originated from plans drawn up by Friedrich Althoff. This "omnipotent" and resourceful director at the Prussian Ministry of Culture perceived at the very beginning of the twentieth century the necessity of modernizing the way science was organized in Germany by providing universities with productive external research institutions. Under excellent working conditions, prominent scholars would be able to pursue their research at one of these institutions, largely unhampered by the obligations of a regular university teacher. The goal was to strengthen and expand the leading position Germany enjoyed in science throughout the world. Althoff had selected as the site for such a "German Oxford" the Prussian domain Dahlem, a sparsely settled suburb on the southern outskirts of Berlin. It took a decade for this idea to become reality, however, and Althoff, who died before his time, did not live to see it. It had demanded new methods of organization and funding: private grants and special attention to the political interests of major industry or finance in science and research also played a central role in its materialization.

Wilhelm II set the initiative for the society with his speech during the centennial celebration of the University of Berlin in October 1910. In it the kaiser

Fig. 2.11. Kaiser Wilhelm Institute of Physical Chemistry and Electrochemistry

proclaimed his intention "to found under my protectorate and name a society which sets as its mission the establishment and maintenance of research institutions." In the following year the Kaiser Wilhelm Society for the Promotion of the Sciences was officially founded. With great pomp and ceremony the first Kaiser Wilhelm Institutes were inaugurated—in the presence of the society's namesake and protector, the German kaiser—in the fall of 1912. The Kaiser Wilhelm Institute of Physical Chemistry and Electrochemistry was among the founding institutes of the Kaiser Wilhelm Society. Its director was Fritz Haber. Like the majority of these institutes at that time, it was located in Dahlem. The management was likewise centered in Berlin—eventually in the general administration building, the seat of its president. From its founding until 1930 the presidency was filled by the theologian Adolf von Harnack, who from the scientific point of view was among the most influential founding fathers of the Kaiser Wilhelm Society. Because Harnack's main office was general director of the Royal Library, or Prussian State Library, he conducted all of his other official business there as well. Until the spring of 1914 the library was located in the *Kommode* in the square by the opera and thereafter in the new library building on Unter den Linden. During the 1920s the society's general administration was moved, along with the official seat of its president, to the palace in Berlin—to the address Palace Portal III.

One frequent visitor to all these locations was Albert Einstein—to settle the details of his appointment as director of the planned Kaiser Wilhelm Institute of Physics or, later on, to discuss that institute's budget and matters of science policy. At first, though, he had much business with Fritz Haber and the Prussian Ministry of Culture. There were difficulties obtaining the necessary funding, and the fact that physics was already the purview of an existing, renowned nonuniversity research institution, the PTR, did not simplify matters. So the founding of a separate physics institute was not very high among the society's priorities, and delays were the inevitable result. Moreover, Haber's Kaiser Wilhelm Institute of Physical Chemistry and Electrochemistry was already addressing problems in the area of physical research. One element of Haber's plan was to extend the scope of research in classical physical chemistry through modern approaches. Among these, Haber counted the new findings on radiation theory and electromechanics, which held promise for the needs of physical chemistry.

Fig. 2.12. View of Dahlem around 1918. The Kaiser Wilhelm Institute of Physical Chemistry is the second large building on the road along the lower edge of the photo.

Particularly as a result of the Solvay Conference in Brussels during the fall of 1911, leading researchers of the day were aware that Planck's quantum of action h played a central part in the interactions between atoms and molecules with electromagnetic radiation.

Haber wanted the young Albert Einstein to solve this fundamental problem at his institute and planned to establish a prestigious research position for him there. To see how realistic these plans were, he took the opportunity of a New Year's vacation in Switzerland in 1912–13 to visit Einstein in Zurich and find out what his career plans for the future were. Upon returning to Berlin he reported to the Ministry of Culture:

> After having turned this idea over in my mind for quite some time, I have become convinced that the realization of this idea would be of the greatest advantage for the Institute, and that, from the personal side, it could probably be attempted with some chance of success. Even though I did not go so far as to give Mr. Einstein any hint of it, I did find out that, completely absorbed by his investigations as he is, he would gladly do without the large course of lectures that he is obliged to give. Further, I ascertained that he has no fundamental misgivings regarding Berlin. (1, vol. 5, trans. p. 327)

Max Planck and Walther Nernst were then drawn into the discussions. Both were among the most reputable and influential representatives of Berlin's scientific community. Both also had connections with high-ranking men of influence in science policy making and industry. Moreover, they both shared a special relationship with Einstein. Planck can certainly be called the "discoverer" of the young Einstein. Very early on, he had drawn attention to the significance of the 1905 theory of relativity and to the genius of its author. Nernst had been enthralled by Einstein's papers on specific heat because they could explain his measurements of the behavior of the specific heat of solids at low temperatures. As a result, Nernst became an early and enthusiastic supporter of Planck's quantum hypothesis. He was also the instigator of the famous Solvay Conference of 1911 in Brussels; the young Einstein was then invited to join the most important physicists of his day at this summit meeting.

The discussions about Einstein by the "triple star" of Berliner physicists, who surely also consulted other local members of their field, soon evolved into an offer more generous than simply a position as department head at Haber's institute—nowadays it might be equivalent to being a scientific member of the Max Planck Society. A well-paid staff membership in the Prussian Academy was to be offered along with the promise of directing a new Kaiser Wilhelm Institute for Physical Research. This part of the plan became possible through the generous support of the banker Leopold Koppel, who had provided a major grant for the founding of Haber's institute as well as for the Kaiser Wilhelm Society in general. Not only was Koppel willing to cover a large portion of Einstein's salary at the academy, but in 1913 he also indicated his willingness in principle to make a substantial amount of capital available for the founding of a Kaiser Wilhelm Institute for Physics. But in October 1913 Einstein had to inform his cousin Elsa: "I haven't heard anything regarding the institute, I don't think about it anymore. It will surely fall through, as it well deserves" (1, vol. 5, trans. p. 357). And three weeks later he wrote her: "The matter of an institute for me has been postponed until after my coming to Berlin. It would, in fact, be good if I were to get some sort of institute; I could then work together with others instead of only by myself. This would be much more to my liking" (1, vol. 5, trans. p. 360).

Einstein's pessimism was certainly not unfounded. Nevertheless, the "mandarins" of Berlin physics, Fritz Haber, Walther Nernst, Max Planck, Heinrich Rubens, and Emil Warburg, submitted a petition in February 1914 to the Prussian Ministry of Culture, the Kaiser Wilhelm Society, and the Koppel Foundation "re. justification for a Kaiser Wilhelm Institute of Physical Research." In it the purpose of the institute was set forth:

> to gather together "at once or one after another particularly suitable groups of researchers in physics to solve important and urgent problems. . . ." The relevant problems would thus be tackled in a systematic manner, both by means of mathematical and physical considerations, as well as, and especially, through experimental analyses conducted in the laboratory by the researchers concerned toward obtaining the most comprehensive results possible. The seat of the institute we would imagine to be in a small building in Dahlem, which would grant the facilities for meetings as well as for the storage of the archive, library, and individual physical apparatus. (21, vol. 1, p. 146)

Regarding the management of the institute, the memorandum envisioned Einstein as the "permanent honorary secretary" of a scientific board of directors to be elected every three years. We might ask whether this structure reveals some hesitation at entrusting the direction of such an institute to such a youthful candidate as Einstein, whose leadership qualities and skill in dealing with tasks related to the selection, financing, and execution of research projects had yet to be seen. Or was it merely in order to minimize costs? Either way, the structure and purpose of the planned institute differed substantially from those of the existing Kaiser Wilhelm Institutes. One could perhaps even conceive the plan as an institutional innovation, because its vision is recognizable today in modern research planning as interdisciplinary or transdisciplinary collaboration, which only appeared in the field of view of science organizers decades later and is currently being practiced by the Wissenschaftskolleg, an institute of advanced study in Berlin, for example.

After these plans had been generally approved by the senate of the Kaiser Wilhelm Society in March 1914 and Haber had written another memorandum on the founding of an institute for theoretical physics, the Koppel Foundation made a binding commitment to cover the construction costs of a modest institute building and an essential portion of its operating costs. But this commitment was attached to the condition that the Prussian state come up with a proportionate sum for the financing and that "the new institute be called upon, in accordance with the existing plan, to work in close cooperation with the already existing two chemical Kaiser Wilhelm Institutes, particularly with the Kaiser Wilhelm Institute of Physical and Electro-Chemistry" (94, p. 199).

When the plans were finally submitted to the Prussian Ministry of Finance in July 1914 to secure one-third of the funding of the project from the Prussian state, the minister rejected it, on the day before the outbreak of World War I. Under the circumstances, the project was no longer deemed practical, and the matter was closed. The Kaiser Wilhelm Society likewise put aside its plans to found any further institutes for the duration of the war.

The fact that soon after arriving in Berlin Einstein moved into an apartment in Dahlem and also received an office at Haber's institute must be regarded as a result of these plans. A personal relationship had developed between Einstein and Haber since their first acquaintance in 1911 at the scientific conference in Karlsruhe. Haber had developed "into an affectionate and ever caring, fatherly friend, which Einstein apparently needed" (76, p. 224). His hospitality to Einstein's wife, Mileva, was part of it. At the beginning of 1914 she stayed in his official villa while searching for a suitable apartment for the family. When the marriage fell apart, Mileva and her sons again found asylum at Haber's; and he was Einstein's companion during the difficult parting at the train station, in the summer of 1914.

Despite, or perhaps because of, their closeness, Einstein found quite a few defects in Haber's character and lifestyle. In December 1913 he wrote his cousin:

> Haber's picture [is] everywhere to be seen. . . . I must unfortunately reconcile myself to the idea that this otherwise so splendid man has succumbed to personal vanity, which is not even of the most tasteful kind, either. This lack of refinement

Fig. 2.13. Fritz Haber and Albert Einstein, 1914

is, unfortunately, just the way of the Berliners. . . . Vanity without real self-esteem. Civilization (nicely brushed teeth, elegant tie, well-groomed moustache, impeccable suit), but no personal culture (coarseness of speech, gestures, voice, emotions). (1, vol. 5, trans. p. 366)

A friendship based on acknowledgment and respect formed all the same between the two in Berlin and even stood the test of World War I. Their reactions to the war were extremely different. During this period Einstein developed into a political scholar and a radical pacifist, whereas Haber placed his talent and energy in the service of the war and became the "father of gas warfare." Although—as Haber wrote Einstein in 1919—"the war years did draw us apart" (1, vol. 9, trans. p. 81), their friendship persisted, and especially during the period of the Weimar Republic Haber was among Einstein's "most special and good-willed" friends and colleagues. In retrospect, at least, it is astonishing that not a single word of public criticism by Einstein about the poison-gas research conducted at Haber's institute is known. They appeared not to have allowed the research at Haber's institute to affect their personal relationship—even though they worked right next door to each other during the first years of the war and saw each other almost daily. The fact that the staff at the institute grew almost tenfold or what they were working on could not have escaped Einstein either.

No information exists about how long Einstein was a guest at Haber's institute with an office of his own. But his Dahlem relations fell away by 1916–17 at the latest, definitely after his move into the city. At that time there were renewed efforts to found a Kaiser Wilhelm Institute for Einstein. In 1917 the winds had suddenly changed, and the shelved plans were drawn out again. The Berlin industrialist Franz Stock offered the Kaiser Wilhelm Society a donation of half a million marks, thereby placing the society in a position to carry the lacking third of the funds that the state had initially been supposed to take up. The founding plans were again plausible—although in a humbler form.

After the senate of the Kaiser Wilhelm Society had dusted off the original plan and reapproved it, the Kaiser Wilhelm Institute of Physical Research was formally christened on October 1, 1917. Announcements about the new institute that Einstein had personally drafted as institute director and had even paid for in advance himself appeared in two Berlin papers. He could easily swallow that expense with his annual compensation of 5,000 marks as director, of course. The board of directors assisting him was composed of Haber, Nernst, Rubens, Warburg, and Planck, the last of whom had accidentally been omitted from the public announcements. The responsibility of the board of directors was "selection of the problems, the methods, and the location of the work. . . . However, suggestions to the board of directors offered by other physicists [were] also to be considered and, in case of approval, the proposed investigations supported" (40, p. 28).

Such a research profile was a novelty for the Kaiser Wilhelm Society, prompting the following commentary by its president, Adolf von Harnack, during a meeting of the membership:

> This institute has a very unique structure differing from all the other institutes of the Society. It has no building of its own and no laboratory of its own, but the means are placed in the hands of a group of appointed physicists; they determine which work is supposed to be performed or which scholars be given grants and which instruments for the furtherance of their investigations be approved. These

On the 1st of October 1917 the

Kaiser Wilhelm Institute of Physical Research

was called into being. Its intended task is to organize and promote a systematic treatment of important and urgent physical problems through the engagement and material support of especially suitable researchers.

The selection of the problems, the methods, and the location of the work lie in the hands of the undersigned board of directors. However, suggestions to the board of directors offered by other physicists are also to be considered and, in case of approval, the proposed investigations supported.

Although the institute will naturally only be able to develop its full potential after the end of the war, the work is supposed to be commenced now. Information about the details should be directed to the cosigning chairman of the board of directors, Professor Einstein (Haberlandstr. no. 5, Schöneberg, Berlin).

The board of directors.
Einstein. Haber. Nernst. Rubens. Warburg.

Am 1. Oktober 1917 ist das

Kaiser-Wilhelm-Institut für physikal. Forschung

ins Leben getreten. Seine Aufgabe soll darin bestehen, die planmäßige Bearbeitung wichtiger und dringlicher physikalischer Probleme durch Gewinnung und materielle Unterstützung besonders geeigneter Forscher zu veranlassen und zu fördern.

Die Auswahl der Probleme, der Methoden sowie des Arbeitsplatzes liegt in der Hand des unterzeichneten Direktoriums. Doch sollen auch von anderen Physikern an das Direktorium gelangende Anregungen von diesem erwogen und die vorgeschlagenen Untersuchungen im Falle der Billigung gefördert werden.

Wenn das Institut auch naturgemäß erst nach Beendigung des Krieges seine volle Wirksamkeit wird entfalten können, so soll doch womöglich schon jetzt mit der Arbeit begonnen werden. Angaben über nähere Einzelheiten sind an den mitunterzeichneten Vorsitzenden des Direktoriums, Professor Einstein (Haberlandstr 5, Berlin-Schöneberg) zu richten.

Das Direktorium.
Einstein. Haber. Nernst. Rubens. Warburg.

Fig. 2.14. Announcement in the *Vossische Zeitung,* December 16, 1917, of the inauguration of the Kaiser Wilhelm Institute of Physical Research

investigations are then conducted at the institutes of the scholars involved; but the instruments remain the property of the central office and are returned to it in order later to be used by other scholars as well. The justifiable hope is thus to strengthen physical research, unite it, and also—to "cheapen" it. (45, p. 50)

The institute's quite respectable budget—75,000 marks per year—did not attain the level of the other Kaiser Wilhelm Society Institutes, but it was still better equipped than most of the scientific facilities at the University of Berlin. Moreover, practically the entire sum could go to the benefit of the planned research projects, because neither institute buildings nor other operating costs needed to be covered. Einstein's preference for attending the institute's business at home (as indicated by the address given in the announcement, Haberlandstrasse no. 5, his private address) also reduced expenses. At the beginning, the secretarial tasks were performed by Ilse Einstein, the eldest daughter of his cousin and lover. Then, after several interim solutions, Helene Dukas assumed this work in 1928.

Einstein seems to have very quickly become disillusioned about his work as director with its bureaucratic and administrative responsibilities. In January 1918 he was already complaining to his friend Michele Besso: "The K. W. Institute involves quite a large amount of correspondence; even so, my correspondence is steadily on the rise" (18, p. 124; 1, vol. 8, trans. pp. 436). And in the summer of 1920 he soberly wrote: "My work is currently not up to much either. I split up my energy, have to deal with immense amounts of correspondence, evaluate, advise, act as protector, but make no progress on the larger questions" (18, p. 152; 1, vol. 10, trans. p. 216).

Although Einstein nominally remained director of the institute until his emigration, in the 1920s he became less and less involved in its daily business. In 1921 Max von Laue was elected to the board of directors, at Einstein's wish, and shortly afterward Laue was named vice-director and increasingly took over the management of the institute's affairs.

A blemish of failure marked not only Einstein's leadership as institute director but also the original governing idea behind the new institute. Nothing came of the plans to draw together researchers to work for a specified period of time on fundamental physics problems, such as quantum theory, or to implement ideas and research projects developed at their own Kaiser Wilhelm institutions. Indeed, no serious attempt to carry out these plans seems ever to have been made. The institute's activities were mostly limited to granting research funds to carefully selected applicants. On average, about a dozen research applications received such support each year, and the approvals were the result of informal conversations among the board members between academy sessions or at relatively unstructured meetings of the directors. For this the institute needed no building of its own.

No distinctive research profile is evident in the approval process either. There were occasionally even problems finding suitable takers for the available funds, resulting at one point in a grant application by the German Entomology Museum being accepted. This flagrantly went against the original intentions of the institute and was in stark contrast to the successes of Arnold Sommerfeld at his university department in Munich or by Niels Bohr at his institute in Copenhagen, for instance. Their cooperative and interdisciplinary research contributed significantly to the development of modern quantum theory. Comparable achievements by Einstein's institute, although originally planned, were not forthcoming. Not even the theory of relativity benefited much; the board deemed only three applications in this area worthy of their support.

As the 1920s progressed, Einstein's institute increasingly lost visibility within the expanding spectrum of national research institutions. The Emergency Association of German Science (Notgemeinschaft), later to evolve into the German Research Association (Deutsche Forschungsgemeinschaft), had been founded in 1920, and the Helmholtz Society had meanwhile likewise established itself as a much better, more comprehensive, and effective promoter of scientific and physical research in Germany. As a result, in the second half of the decade, the conceptual basis of Einstein's institute began increasingly to be regarded as out of date, and new organizational forms were proposed to turn it into a proper institution for physical research. But none of these plans could overcome the economic restraints of the time. Not even the intention by the American Rockefeller Foundation to grant major support for these plans could change things. Its materialization had to await the Third Reich, when a modern physics building was erected in Dahlem at Boltzmannstrasse nos. 18–20. The Kaiser Wilhelm Institute of Physics was thus, in fact, refounded under the directorship of Peter Debye.

Site 10. Imperial Institute of Physics and Technology
Physikalisch-Technische Reichsanstalt (PTR)
Currently the Physikalisch-Technische Bundesanstalt
Institut Berlin
Abbestrasse nos. 2–12
10587 Berlin (Charlottenburg)

subway stop Ernst-Reuter-Platz (U2)
from there 5 min. by foot

"Warburg wanted to juggle me into the Reichsanstalt in Berlin," Einstein wrote in a letter to his friend Zangger in Zurich in June 1912 (1, vol. 5, trans. p. 307). But these initial plans to get Einstein to come to Berlin fell through. The person responsible for this failure was probably Einstein himself, who could not regard the position of "in-house theoretician" offered him at the PTR as particularly attractive. He had just recently accepted a well-paid regular professorship in theoretical physics at the renowned Federal Polytechnic Institute in Zurich. Besides, international recognition of his work was steadily rising. He had been counted among the leading physicists invited to the exclusive Solvay Conference in Brussels in the fall of 1911 to discuss the issues of the day, particularly the problem of quanta. On that occasion he made the acquaintance of Emil Warburg, whose experimental confirmation of Einstein's basic photochemical law in papers he had written on photochemical energy conversion was already reason enough for the two men to be interested in each other. Warburg had carried these measurements out at his home institution, the PTR, as its president since 1905. The PTR had been founded with the substantial support of Werner von Siemens in 1887 as the premiere metrological testing institution and foremost research establishment of the German Reich. It was located in Charlottenburg, at that time still a separate town. Besides fulfilling its immediate purpose, it was equally successful in establishing its activities on a much broader basis. Following Siemens's maxim that in the long run solid scientific (basic) research is the basis of technological and economic success, the PTR's agenda complemented the treatment of current problems in technology and metrology with fundamental research and precision physics. As a result, the PTR was not just the largest physics institute in the world at the turn of the century with its exemplarily equipped laboratories but also the international leader in the field of precision physics and technology. Its famous accomplishments included the measurements of radiative heat from "black bodies," which served as the basis for Max Planck's formulation of his radiation law with the quantum hypothesis in the fall of 1900.

Under Warburg's presidency the PTR underwent a process of modernization, opening laboratories devoted to low-temperature physics, radioactivity, and, of course, photochemistry. Warburg's plan to open the PTR's research to modern physics, and not least to problems of quantum physics, clearly included "juggling" the young Einstein into the PTR—just as he had enlisted other promising young physicists into the institute's service, such as Walther Bothe, Hans Geiger, and Walther Meissner.

As we have seen, this plan fell through, ending instead in the idea of luring Einstein to Berlin with a more attractive offer. When Einstein was eventually safely installed as full-time staff scientist of the Berlin academy, his good relations with Emil Warburg and the PTR in general did not end there. During his first visits to Berlin, Einstein stayed not only with his relatives but for a while also at Warburg's presidential villa. Warburg also repeatedly visited Einstein, such as in 1916–17, when Einstein was seriously ill. Later he was among the guests invited to the occasional soirées held in the Einstein home on Haberlandstrasse. During World War I, Einstein arranged for Warburg's son Otto, "one of

Fig. 2.15. Model of the Physikalisch-Technische Reichsanstalt (PTR), 1887

Germany's most talented younger biologists with great promise" (1, vol. 8, trans. p. 512), and future Nobel laureate, to be withdrawn from the battlefield.

Warburg evidently invited Einstein to work as a guest scientist at the PTR soon after his arrival in Berlin. This invitation is explained by the fact that Einstein was concentrating on other things besides the final design and formulation of the general theory of relativity. He was also working on the experimental verification of an almost century-old problem—proof that molecular currents are the cause of magnetism, as postulated by the French physicist André Marie Ampère. This hypothesis had never been experimentally confirmed, and Einstein himself had had little success with experiments he and his friends had conducted while he was still living in Berne.

In Berlin he resumed these efforts in hope of better luck in the PTR's excellently equipped laboratories with staff scientists specializing in the necessary precision measurements. He carried out the appropriate experiments during the winter of 1914–15 as a "guest scientist" of the PTR along with the Dutch physicist Wander Johannes de Haas, a talented researcher in the field of magnetism who had been working there as a staff scientist since the beginning of 1914. The records do not reveal specifically in which laboratory these experiments were performed, but the main building is a likely location. The majority of the laboratories of the second department for electricity and magnetism, in which Einstein and de Haas were registered as guest employees, were housed there.

Fig. 2.16. The main building of the PTR, in which the laboratories of the department for electricity and magnetism were located, 1920s

The basic idea behind Einstein and de Haas's experiments was the assumption that the magnetic moment of a circular current is proportional to the mechanical angular momentum of the electrons generating the current. If a cylindrical rod of soft iron is suspended on a thin fiber of quartz inside a coil and its magnetic moment is altered by periodic magnetizations of the coil, according to the law of conservation of angular momentum the iron rod must start rotating. Simple though the basic apparatus seems to be, it was very difficult to perform accurately and required a knack for experimentation to be able to separate the expected effect from the many perturbing influences.

Einstein wrote his friend Michele Besso about this in February 1915: "A wonderful experiment; what a pity that you can't see it. And how tricky nature is, when you want to deal with it experimentally! Experimenting is becoming a passion of mine, even at my ripe old age" (1, vol. 8, trans. pp. 68).

Einstein reported on their successful results soon afterward, on February 19, 1915, at the colloquium of the Physical Society, and other talks on the same topic were delivered in the next few months. It soon became apparent, however, that although successful in principle, their determination of the gyromagnetic constant, the relation between the change in angular momentum and the magnetization, was far too imprecise. So the claimed agreement between theory and experiment arose more out of theoretical bias than out of the measurements obtained. The correct explanation had to await the discovery of electron spin in 1925, which has a decisive effect on ferromagnetism. All the same, Einstein and de Haas did succeed in providing the first (qualitative) confirmation of Ampère's hypothesis of molecular currents.

Einstein's relationship with the PTR was not restricted to being a guest on its premises. In December 1916 he was appointed "by supreme decree" of Kaiser Wilhelm II as a member of its board of trustees. Einstein became the successor to Ernst Dorn, the physicist from Halle who had died that June. In the fall, initial agreement had been reached between the Ministry of the Interior and the president of the PTR, whereupon the issue of Einstein's Swiss citizenship was discussed:

> The statutes of the Imperial Institute do not forbid the appointment of foreigners to the Board of Trustees, nevertheless installations at the institution made for experiments in the fields of military and marine technology are being kept secret. In view of the fact that Einstein has been appointed to the Royal Academy of Sciences, I believe I may also grant his appointment to the Board of Trustees as advocated by the Reich Bureau's president. (21, vol. 1, p. 158)

In the nomination petition to the emperor, however, the focus is entirely on Einstein's prominent scientific achievements:

> In fact he is, without a doubt, one of the most sharp-witted and original theoretical physicists alive; the theories he developed in pioneering papers serve numerous experimenters inside and outside the country as the grounds and guidelines of their research. He has also worked as an experimenter and, in particular, just recently at the Imperial Institute produced in a highly important paper the experimental proof for the existence of Ampère's molecular currents in magnets. He is also interested in practical problems, so there is particularly much promise in his collaboration on the work at the Reich Bureau. (21, vol. 1, p. 159)

The board of trustees was something like a supervisory board of experts on scientific and technical work conducted at the PTR. According to its rules of procedure, its tasks primarily included consultations on the overall working agenda but also on such issues as the budget, general personnel policy, the hiring or appointment of higher civil servants, and approvals of visiting scientists. The members of this board included leading scientists and engineers, such as Röntgen, Planck, Nernst, and Carl von Linde, as well as representatives of the state and industry. No term limitations were provided for by the statutes so, as a rule, one had a seat and a vote on this expert committee for life or the duration of one's professional activity.

The board met once a year in the spring, so its supervisory function over the activities of the PTR could naturally only be very superficial. Einstein attended these trustee meetings, which usually lasted three days, relatively regularly—it was only at the end of his Berlin period that his interest and enthusiasm in this capacity waned. According to the records, the last two trustee meetings he attended were in 1927 and 1930. On the latter occasion he was apparently just a mute listener; at least there is no record of any of his comments in the proceedings. It was very different in the early days. Then Einstein used to interject many pointed remarks during the board's discussions. Famous as he is to us now as the theoretical physicist par excellence, he then presented a more multifaceted side, often attempting to influence the experimental work under way at the PTR with concrete suggestions. He also expressed his opinion on various issues of science policy and management—about the right of PTR employees to take out patents, for instance.

In the year 1921–22 Einstein even seems to have become a de facto visiting scientist at the bureau again. Concrete evidence of this is lacking in the official files, but Einstein's correspondence provides some clues. In the summer of 1921 he had "thought of a very interesting and fairly simple experiment on the nature of the emission of light" (16, trans. p. 56), with which he hoped to settle the then very lively debate about whether light was wavelike or particle-like in character. To carry out this highly sensitive experiment, which was supposed to examine the character of the light emitted by canal-ray particles, he again secured the proven competency and exceptional apparatus available at the Reich Bureau. This time his congenial partners were Hans Geiger and Walther Bothe from the radioactivity laboratory. With their "splendid cooperation" the "interesting experiment" could soon be realized. It was—as he wrote to Max Born—"my most impressive scientific experience in years" (16, trans. p. 65).

But his optimism did not last very long. Just a few weeks later he was forced to admit: "I too committed a monumental blunder some time ago (my experiment on the emission of light with positive rays), but one must not take it too seriously. Death alone can save one from making blunders" (16, trans. p. 71).

The arrival of the Nazis in government also had consequences for Einstein's involvement in the PTR. It was, after all, a state institution, so politics had a direct influence on its affairs. By notice of the Reich Ministry of the Interior, the

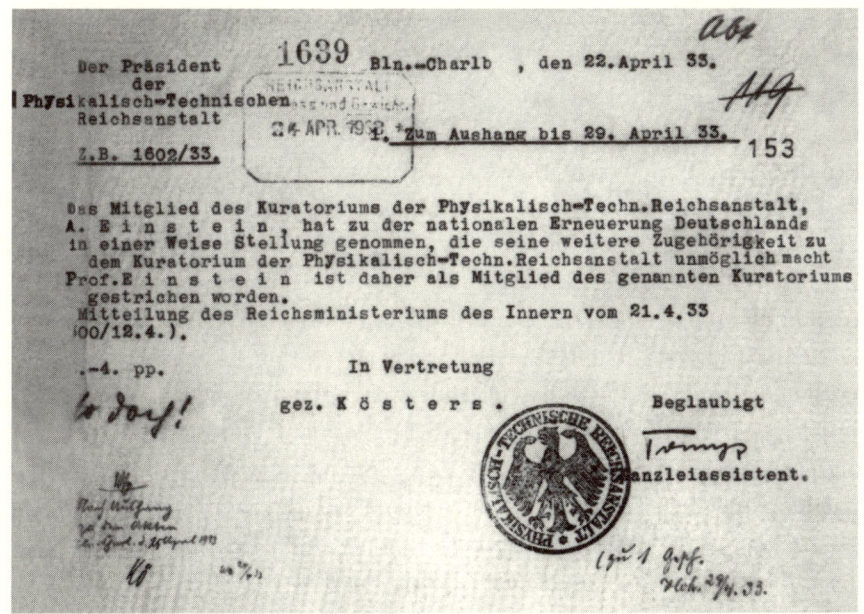

Fig. 2.17. Facsimile on the removal of Albert Einstein from the PTR's board of trustees

authority in charge, it was decreed in April that Einstein's name be struck from the membership list of its board of trustees because he "has taken a stance about the national renewal of Germany that makes impossible his continued membership on the Board of Trustees of the Reich Imperial Institute of Physics and Technology" (54, p. 101).

Thus ended Einstein's relations with the PTR, and they were not resumed when the institute was reopened in Braunschweig as the Physikalisch Technische Bundesanstalt (the Federal Bureau of Standards) after World War II. The abrupt closing of the "Einstein case" was not a shining moment in the annals of an otherwise so successful institution.

Site 11. AEG Research Laboratory
Now the Anthropolis retail center
Holländerstrasse nos. 31–34
13407 Berlin (Reinickendorf)

subway stop Franz-Neumann-Platz (U8),
from there 5 min. by foot

It is common knowledge that Albert Einstein worked for many years as a patent official. The fact that he held a number patents of his own and was a quite competent inventor himself is not so commonly known. Nevertheless, his inventions were not rare exceptions to his scientific activity, and the two spheres of his accomplishments are not unrelated. In his memoirs from 1947 he notes that his "work on the final formulation of technical patents [was] a true blessing for me. It forced me to think in a multifaceted way, also giving me important inspiration in thinking about physics" (20, p. 12).

Although Einstein was constantly inventing things throughout his life (61c), his Berlin years mark a high point in this activity. Of some fifty patents Einstein registered, usually jointly with a colleague, well over half of them date to his Berlin period. There he came into contact with Hermann Anschütz-Kaempfe, an engineer from Kiel. Einstein advised him about a patent dispute involving his invention, the gyroscopic compass and worked with the engineer to improve its design. The result of this collaboration was some jointly held patents and a considerable amount of supplementary earnings for Einstein.

A perhaps closer working relationship with a fellow inventor was with Rudolf Goldschmidt, who had conceived some important inventions in the area of

Fig. 2.18. The AEG research laboratory building, around 1930

wireless telegraphy before World War I. The mechanical transmitter he developed attracted particular attention. When tube technology made Goldschmidt's invention more or less obsolete, he turned his attention to medical technology in the 1920s, and Einstein advised him on various devices he developed—for heart examinations and for measuring pulse rate, for example. At the end of 1928 the two even registered a patent together for a new electromagnetic driving mechanism for loudspeakers. In celebration of the successful filing of their patent, Einstein sent his partner a portrait with the following humorous verse at the bottom:

> A little tech now and then
> even brooders can amuse.
> Boldly do I think ahead:
> Let's twosome lay another egg.

> Ein bisschen Technik dann und wann
> Auch Grübler amüsieren kann.
> Denn kühnlich denk ich schon so weit:
> Wir legen noch ein Ei zu zweit.

Goldschmidt promptly retaliated:

> Let's twosome lay an egg, you say?
> That might present a quandary!
> The best solution to the task
> is perhaps, if you agree,
> let's lay the eggs in rivalry
> and omelets be our recipe.

> Das Legen eines Eis zu zweit
> Das bietet manche Schwierigkeit!
> Die beste Lösung vom Problem
> Wär', wenn sie Ihnen angenehm,
> Wir legen Eier um die Wette
> Und fabrizieren Omelette.

How this egg-laying contest between the two inventors was actually carried out is no longer reconstructable. In any case, in 1931 they joined forces again to help out an acquaintance of theirs. The singer Olga Eisner was suffering from a gradually worsening loss of hearing. So they invented an electro-acoustical hearing aid for her. The basic idea of converting an acoustical signal into electrical oscillations in order to amplify it came from Einstein. The signal was supposed to be transmitted by some sort of membrane directly to the skull so that the bone could conduct it on to the hearing organ. Goldschmidt's role in laying this "omelet" was to transform the idea into a technically feasible hearing aid. Some important progress was made, but the device wasn't ready for patenting before the two had to flee Germany in 1933. At the beginning of the 1940s Goldschmidt filed for another patent in England based on their earlier work together and invited Einstein to continue their collaboration on it. But Einstein could not, as he put it in his letter to Goldschmidt, "resolve to go on an new escapade into the realm of technology. First of all, my own business takes up all of my time and, second, I know from experience that any activity beyond my field only draws an ugly 'publicity' with it that definitely must be avoided" (12, AE to R. Goldschmidt, December 20, 1941). In any case, the development of electronic hearing aids soon overshadowed Einstein's and Goldschmidt's idea for an electro-acoustical hearing aid.

An equally short-lived inventive idea of Einstein's that he tried to materialize toward the end of the 1920s managed to catch the interest of a leading German

Fig. 2.19. Portrait of Albert Einstein with his dedication for Rudolf Goldschmidt, 1928

electrical combine, AEG, in Berlin—along with other firms as well. An article about a tragic accident was its inspiration. A family reportedly had suffocated in their sleep in the toxic fumes of their leaking icebox. Starting from the assumption that the refrigerant had escaped from a defective or leaking compressor or pump, Einstein thought about how to construct a safely closed refrigerant cycle without any mechanical pumps or similar drive assemblies with moving components requiring sealing. To find the solution he was relied on the help of a young Hungarian physicist.

Leo Szilard, born in 1898 in Budapest, came from the horde of "unsettlingly intelligent Hungarians" spilling out of their homeland after World War I to finish off their scientific education in Germany, particularly in Berlin. Szilard first studied engineering at the Polytechnic Institute in Charlottenburg (Technische Hochschule) before turning to physics. In 1922 he took his doctorate under Max von Laue at the University of Berlin with a thesis on thermodynamics. This thesis was what brought him into contact with Einstein, who strongly supported it and recommended he first become a patent examiner, arguing that the usual pressures of an academic career with its demand for "golden eggs" was not necessarily beneficial for the development of a young scientist. Szilard did not follow his friend's paternal advice but earned his postdoctoral degree four years later with a habilitation thesis that was to become one of the pedestals of modern information theory. Until his emigration in 1933, he continued working at the University of Berlin as a private lecturer.

Einstein's advice seems not to have been entirely without consequence, though, because Szilard continued to grapple with technical problems in parallel with his scientific work. Many of his patents date from 1923 on—probably not least as a means to bridge the economic uncertainties of an academic career, especially at the stage of unsalaried private lecturer. In 1926 Einstein started supporting Szilard's work as an inventor when they collaborated on the development of "small refrigeration machines for households." Their first patent for such a device was filed in October 1926, to be followed by others. On the hope of making some financial profit from these inventions and patents, they decided how to divide the credit well in advance. They agreed to regard every invention as their common property and to divide the profits in equal shares unless Szilard's income was lower than that of a university assistant, in which case his share would be larger.

Most refrigerators of the time worked with motor-driven compressors, which sucked up the refrigerant and condensed it. Einstein and Szilard's basic idea was to develop a refrigerating device without such trouble-prone moving parts. The initial intention was to develop a refrigerator that functioned on the basis of the principle of absorption. Their prototype quickly passed the laboratory testing stage. So a contract was signed with the firm Bamang-Meguin, which operated the gasworks in Berlin and Sachsen-Anhalt. However, a year later the company got into financial difficulties and terminated the majority of its development projects, including the Einstein-Szilard absorption refrigerator. A new partner was found in the Swedish firm Electrolux, which agreed to buy the patents from the two inventors for a negligible sum. But then the company lost interest in developing their invention, preferring to concentrate on their own designs, and so the patents disappeared into the company safe.

They had more luck at first with another collaboration they had also started in 1927 with Citogel-Gesellschaft. This chemical and technical products firm in Hamburg took to their idea of an evaporation refrigerator: the pressure in a small chamber is lowered by means of a water-jet blast, causing a water-methanol mixture to evaporate. This effects a sudden cooling of the surrounding area from which the evaporation heat had been extracted. The water-jet blast also removes the refrigerant fumes, but unfortunately the methanol is lost in the process, necessitating constant refilling. The technical feasibility of this design was first tested in the laboratory of the Polytechnic and at the workshop of a machine factory in Berlin (probably Bamang at the address Reuchlinstrasse nos. 10–17 in the district of Moabit), and it worked surprisingly well. At the spring fair in Leipzig in March 1928 Citogel was even able to present its "people's fridge" (*Volks-Kühlschrank*). This presentation must have raised hopes, because after the fair the common capital stock of the company rose by 50 percent. Einstein and Szilard also continued to work on perfecting their evaporation refrigerator. In May 1928 Einstein wrote to his son Hans Albert: "Szilard and I have filed a patent for something very pretty, a heat insulator for refrigerators. Cork tiles are normally used. We use a few parallel paper walls that are distanced. At least the same insulation is achieved without incurring any mentionable cost" (12, AE to H. A. Einstein, May 12, 1928).

Half a year later the situation didn't look so rosy anymore. There were many complaints by users of the model. Szilard reported "not just that methanol turned out to be unexpectedly expensive in retail sales; most importantly, the sophisticated cooler depended on a constant water pressure. But that varied so much from building to building and from one floor to the next that the invention never did get on the market in the end" (41, p. 98).

Einstein and Szilard's conception of a "people's fridge" based on the principle of evaporation was both original and unique. Even so, it had no impact on further advances in refrigerator technology. Another of their insights had more influence and success. As they were developing their evaporation prototype, they also pursued a parallel idea of constructing a refrigerating engine using an electromagnetic unit. Einstein once outlined this idea as follows: "an alternating current produces an alternating magnetic field that keeps a fluid potassium-sodium alloy moving. The fluid alloy performs an alternating motion inside a closed housing and acts as a piston of a pump for the refrigerant, which is thus mechanically condensed and upon reevaporation generates the cold" (46, p. 289).

At the beginning of 1928 they approached AEG to find out whether it was interested in producing such a pump for refrigeration purposes. Einstein's personal connections with the company management helped bring about a contract with the two inventors in the fall of 1928.

The necessary facilities for the developmental work were made available in the company's private research institute in January 1929. AEG not only paid the patent licenses due to the two inventors but also assumed the cost of a team of three researchers: Szilard and two engineers. As head of the laboratory, Szilard received a monthly salary of 500 marks, which was considered a good income, besides the not insignificant patent license payments.

The AEG research institute was located on Holländerstrasse at the corner of Aroser Allee in a building dating from 1919 that also accommodated the company's training workshop. The research institute had just recently been founded in 1928 and was being directed by Carl Ramsauer, a physicist of international renown who made the institute into one of Germany's most productive indus-

Fig. 2.20. Prototype of the Citogel "people's fridge," around 1928

DEUTSCHES REICH

**AUSGEGEBEN AM
30. MAI 1933**

REICHSPATENTAMT
PATENTSCHRIFT
№ 563403
KLASSE 17a GRUPPE 3₀₁
S 82663 I/17a
Tag der Bekanntmachung über die Erteilung des Patents: 20. Oktober 1932

Dr. Leo Szilard in Berlin-Dahlem und Dr. Albert Einstein in Berlin
Kältemaschine
Patentiert im Deutschen Reiche vom 13. November 1927 ab

Die Erfindung betrifft eine Kältemaschine, bei welcher der Dampf eines Kältemittels durch ein flüssiges Metall verdichtet wird. Die Bewegung des Metalls erfolgt hierbei dadurch, daß ein Magnetfeld auf das stromdurchflossene Metall einwirkt. Das Metall befindet sich dabei vorzugsweise in einem schmalen Spalt. Gegenstand der Erfindung ist eine besondere Art der Ausbildung der Vorrichtung, mit deren Hilfe das flüssige Metall in Bewegung gehalten wird und die im besonderen auch gestattet, die Strömungsrichtung des Metalls intermittierend umzukehren.

Fig. 1 zeigt ein Ausführungsbeispiel der Erfindung im Schema gezeichnet. 1 ist eine Vorrichtung, in welcher das Feld eines Elektromagneten auf das Quecksilber einwirkt, durch welches mit Hilfe der Elektroden 2 und 3 ein elektrischer Strom hindurchgeschickt wird. Bei Verwendung von Wechselstrom sind Strom und Magnetfeld möglichst in Phase gehalten, und es tritt dann im Quecksilber eine Kraftwirkung auf, die es aus dem Zylinder 4 in den Zylinder 5 hineinpreßt und bei Umpolung der Elektroden umgekehrt aus dem Zylinder 5 in den Zylinder 4 befördert. Die Umpolung erfolgt selbsttätig mit Hilfe der Kontakte 6 und 7, die in den seitlichen Ansatzrohren 8 bzw. 9 untergebracht sind und bei Berührung mit dem Quecksilberspiegel in diesen Ansatzrohren je einen Hilfsstromkreis schließen. Die genannten Ansatzrohre sind durch dünne Leitungen 10 bzw. 11 mit den Zylindern 4 bzw. 5 in Verbindung, durch welche das Quecksilber in sie eindringt, wenn der Quecksilberspiegel im entsprechenden Zylinder hochgestiegen ist, und durch welche das Quecksilber aus dem Ansatz herausfließt, wenn der Quecksilberspiegel im betreffenden Zylinder heruntergesunken ist. Bei dieser Anordnung ist die Lage des Quecksilberspiegels im Ansatzrohr keine eindeutige Funktion der Lage des Quecksilberspiegels im Zylinder. Die Verhältnisse liegen vielmehr so, daß, wenn die große Masse des Quecksilbers zwischen den beiden Zylindern 4 und 5 hin und her pendelt, das Quecksilber in den Ansatzröhren 8 und 9 mit großer Phasenverschiebung folgt. Trifft man die Anordnung so, daß durch den Kontakt 7 beim Schließen des betreffenden Hilfsstromkreises (wenn also das Quecksilber im Zylinder 5 hochgestiegen ist) der elektrische Strom im Stromkreise der Elektroden 2 und 3 umgekehrt und damit zugleich die Kraft, welche das Quecksilber zwischen den beiden Zylindern hin und her treibt, umgekehrt wird, so wird nach Stromschluß des Kontaktes 7 das Quecksilber aus dem Zylinder 5 angesaugt und in den Zylinder 4 hineingedrückt; obwohl der Quecksilberspiegel im Zylinder 5 sofort nach Stromumkehr herunter zu sinken beginnt, steigt entsprechend der genannten Phasenverschiebung im Ansatzrohre 9 der Quecksilberspiegel noch eine Zeitlang an (so lange, bis die beiden Spiegel gleich hoch stehen), und der Hilfsstrom-

---

Fig. 2.21. Patent specification for the refrigerator design by Albert Einstein and Leo Szilard, 1927

trial laboratories. Almost two hundred scientists and technicians worked there on a broad spectrum of industrially relevant research. Szilard's development team could hardly have taken more than a marginal place at the institute.

Although the lion's share of the developmental work on the electromagnetic pump was executed by Szilard and his two coworkers, Einstein was certainly not merely a silent partner. Recollections document that he was a regular visitor at the research laboratory in Reinickendorf, and frequent discussions also took place in Einstein's apartment on Haberlandstrasse. Within a few months a pilot model for the Einstein-Szilard pump was ready, but it proved faulty. The noise generated by the pump was so loud that it "howled like a jackal." After the worst defects had been eliminated in 1931, the next stage was building a prototype and comparing its efficiency against the refrigerators then in use. Competitive results were achieved, but that was no guarantee that the new model would survive on the market. The introduction of Freon made the conventional models much less hazardous and even more efficient to operate. Affected by the economic crash, AEG was forced to put its research and development projects on the back burner. Szilard's team was granted a year's respite before its research on electromagnetic-pump refrigerators was finally terminated in the summer of 1932.

The foresightful Szilard had had a premonition about how politics in Germany might develop two years earlier. He then prophetically wrote to Einstein: "If my nose isn't misleading me, week by week new symptoms [indicate] that one cannot count on a peaceful development in Europe within the next 10 years. . . . I don't even know whether we'll succeed in finishing the design of our refrigeration machine in Europe" (41, p. 99).

Szilard was to prove right on all points, because a few months after the end of their collaboration with AEG the National Socialists came to power, driving Einstein and Szilard out of Germany. They made various attempts to find a new sponsor among American and British companies for their induction-pump refrigerator but without success. Although their partnership had ended, the two inventors continued to stay in touch in American exile. In the summer of 1939, Szilard came with his countryman and colleague Eugen Wigner to see Einstein in his summer vacation home on Long Island. Their purpose was to convince him to sign a letter addressed to the American president, Franklin D. Roosevelt, warning about the possibility that Germany was developing an atomic bomb. This letter was an influential impetus—but by no means the only one—leading to the initiation in 1942 of what became known as the Manhattan Project, the United States' intense effort to build its own nuclear bomb.

Fig. 2.22. Einstein and Szilard signing their letter to the American president (a reenactment after 1945)

The collaboration between Albert Einstein and Leo Szilard had another repercussion for nuclear technology as well. While working for the Manhattan Project on Enrico Fermi's reactor experiments, Szilard suggested using the principle underlying the Einstein-Szilard pump for the reactor's cooling cycle. Such induction pumps find many applications in the heat exchangers of modern reactor facilities by virtue of their high operational safety.

Site 12. Archenhold Observatory
Alt-Treptow 1
12435 Berlin (Treptow)

met. train station Treptower Park (Ring)
   or Plänterwald (S8, S9),
from either, 10 min. by foot

Publicity and popularizations of advances in science and technology were not taboo to scholars like Einstein who strove to make the results of their research accessible to broader audiences. There is an abundance of popular essays and newspaper articles by him on the subject of his investigations in physics. The book he wrote with collaborator Leopold Infeld in American exile without a doubt marks the height of his activity as a science popularizer. *The Evolution of Physics* sketches in profound yet simple language the basic trends in the development of physical theory and therefore the very foundations of modern physics. But books like that, by the most important scientist of the day, were not at all common at the time, not even for Einstein. Just three decades before, he had quite brusquely refused a publisher's invitation to write a popular essay on relativity theory, saying, "I cannot imagine how this subject can be made accessible to a wider audience. Comprehension of this subject requires a certain schooling in abstract thinking, which most people do not acquire because they do not need to" (1, vol. 5, trans. p. 128).

Once in Berlin, Einstein gradually moved away from this standpoint generally held by his fellow academicians when he became aware of his rising popularity with the general public. The Berlin dailies and weeklies as well as foreign periodicals began to receive reports by him. Einstein had hardly settled into his new city of residence when the *Vossische Zeitung* contacted him about writing an article on relativity theory, and Einstein complied. The article appeared on April 26, 1914.

Fig. 2.23. Archenhold Observatory in Treptow, Berlin, around 1920

This article may have inspired another of Berlin's great "popularizers," the astronomer Friedrich Simon Archenhold, to invite Einstein to present a talk at the public observatory he had founded in Treptow. This observatory, which now bears its founder's name, had been established in 1909 with public funds in the wake of the labor union movement. Its purpose was to bring knowledge about the stars closer to larger segments of the population and to instill a fascination for the universe and science in general. In furtherance of these educational goals, astronomy courses were offered, along with much journalistic publicity and exhibitions. Regular public lectures on popular science were also held in the observatory's halls. By virtue of his good connections and reputation as a scientist, Archenhold succeeded in attracting famous scholars to these events. Speakers in the Treptow lecture series included the American astronomer Percival Lowell, the polar scientists Roald Amundsen and Fridtjof Nansen, the geophysicist Alfred Wegener, and Albert Einstein.

Einstein's lecture on June 2, 1915, "on the relativity of motion and gravitation" attracted the attention of "a relatively large number of listeners," to quote a report in the *Vossische Zeitung*. It was one of the first of Einstein's lectures to inform a general audience about current research results pertaining to the general theory of relativity. It was the very first lecture on the topic in Berlin. So it is particularly noteworthy because it offers a glimpse into "the researcher's workshop," so to speak. Einstein was able to bring his work on relativity to completion only a few months later, in the fall of 1915.

For Archenhold this lecture was probably the pinnacle of his own lifelong work on relativity theory. He had authored a few brief articles on the subject even before 1915, and Einstein's talk was followed up by others delivered by largely unknown scientists in the years that ensued. The observatory's periodical *Das Weltall* published three longer articles on relativity theory as well. When the theory and its author became the brunt of public slander and anti-Semitic insults in the 1920s, Archenhold felt honor bound to make the facts about the scientific foundations and consequences of Einstein's theory generally known and to influence public opinion with lectures and publications at the observatory. So it is not surprising that Einstein would regard Archenhold as a "welcome" contemporary. All the same, he would not let himself be completely overwhelmed by him either. When Archenhold invited him in 1926 to take part in the observatory's upcoming Mars exhibition, Einstein declined with the friendly but definite words that he did not want to serve "everywhere as a symbolic bellwether and nimbus" (52, p. 330).

Fig. 2.24. Friedrich Simon Archenhold, 1931

The young writer Anna Seghers ought to have received the same treatment when she visited Einstein in Caputh in 1931 to invite him to give a lecture at the Marxist Workers School (MASCH), of which her husband, László Radványi, was the director.

> I was convinced that Einstein would accept the assignment. Why shouldn't he? He was clever, he was for new, progressive things, he too had been pestered by the reactionaries.
>
> With the verve that certainty gives, particularly to a young person who has not yet experienced much contradiction, I told him about the MASCH. . . . Einstein listened very attentively. A school in which people from the factories, the unemployed, and all who had not had an opportunity to learn about things of importance, are explained the laws of life, the essence of science and art! He thought about it, he nodded. His wife interjected, worried as every wife is: "You have to decline! You did resolve not to accept any more talks." Einstein said: "This is an entirely different kind of talk. It interests me." (75, p. 15)

The talk took place on October 26, 1931, as the inaugural event of the new school year in the auditorium of the MASCH on Weinmeisterstrasse in the center of town. The title was "What the Worker Should Know about Relativity Theory." As the talk was not published and the manuscript has not survived, we don't know what specifically Einstein tried to convey to Berlin's factory workers about his theory of relativity. But the talk attracted others besides its target audience. Left-wing personalities had also come to listen to it, including the composer Hanns Eisler and the writer Bertolt Brecht. Brecht erroneously dates the talk to 1930, however, and specifies the scholar's topic as causality. Einstein did not deliver two talks at the MASCH, so Brecht must have misdated it. But it is possible that Brecht attended a different public talk. Einstein offered an entire cycle of lectures at the Volkshochschule, Berlin's adult education center, on "basic tenets of motion and the equilibrium of bodies" at the beginning of 1920.

Fig. 2.25. Title page of the MASCH course catalog for the school year 1931–32 with the announcement of Einstein's talk "What the Worker Should Know about Relativity Theory"

This cooperation may have continued between Einstein and this or another educational establishment of the Berlin workers association. In any case, in a letter to Heinrich Zangger, Einstein reported: "There is evidently enormous interest among these people" (1, vol. 9, trans. p. 205).

Site 13. Einstein Tower
Telegraphenberg
Albert-Einstein-Strasse
14473 Potsdam

By met. train (S7) or regional express
    lines to Potsdam main station
from there, 15 min. by foot, uphill

When the solar eclipse expedition led by the British astronomer Arthur Stanley Eddington in 1919 confirmed that gravity deflects light, one of the predictions of Einstein's general theory of relativity, the theory and its author suddenly became the focus of public attention. Another fact also became glaringly obvious: the general theory of relativity, unlike virtually any other physical theory, was the work of a single individual, and few had wanted to follow Einstein's lead from the special theory to general relativity. Einstein repeatedly complained about the general disinterest in his work. At the beginning of 1914 in a letter to his friend Michele Besso he grumbled about "the fraternity of physicists [behaving] rather passively with respect to my gravitation paper" (1, vol. 5, trans. p. 374).

One result of this lack of interest in or open-mindedness toward his work on generalizing the special theory of relativity was that virtually no one in Germany gave any serious thought to supporting any attempts to test the consequences of his theory experimentally. To his colleague in Munich, Arnold Sommerfeld, Einstein complained: "Only the intrigues of pitiful persons prevent this last important test of the theory from being carried out" (1, vol. 8, trans. p. 153).

This "last important test" was the one that the Englishman Eddington successfully performed four years later when he measured the deflection of light rays traveling in the vicinity of larger stellar masses. Not all German scientists were wearing "blinders," of course; at least one astronomer took an early interest in the astronomical consequences of the general theory of relativity. But he, Erwin Finlay-Freundlich, like Einstein, was a professional loner and merely a subordinate assistant at the University of Berlin's observatory, which was located

Fig. 2.26. View of Telegraph Hill with the Einstein Tower in the foreground on the left, 1936

in Babelsberg, Potsdam. When Einstein sent an inquiry to the observatory in 1911, Freundlich started working on testing the astronomical consequences of the general theory of relativity, which was then still at an early stage of its development. But his position prevented him from devoting himself entirely to Einstein's ideas; he was supposed to be taking orders from his superior, Karl Hermann Struve. That meant taking part in the ongoing research at the observatory, which centered on photometric observations and determinations of star positions. These routine tasks were typical of the kind of research being conducted at German observatories, so Einstein's novel and revolutionary theory of gravitation generally fell on deaf ears among the German community of astronomers. It seemed far too bold, especially considering that positional astronomy, which used Newton's classical theory of gravity, could do very well without it. These doubting astronomers were in good company. Even the theoretical physicist Max Planck, Einstein's personal supporter, who really was qualified to make such judgments, kept his distance from Einstein's bold line of reasoning. When Einstein was being welcomed into the academy in 1914, Planck referred vaguely to his dangerous tendency of "occasionally wandering too far off into obscure areas" (64, vol. 2, p. 247).

Unlike the majority of his fellow astronomers and, specifically, contrary to his boss's wishes, Freundlich devoted his "hours of leisure"—as he put it in his curriculum vitae—to the astronomical consequences of Einstein's research on the general theory of relativity. The initial focus was on demonstrating that the sun's gravitational field deflects passing light. To this end, Freundlich systematically searched through the available observational data from earlier solar eclipses. Neither these efforts nor a solar eclipse expedition that Freundlich and his team undertook to the Crimea in August 1914 were successful. The outbreak of World War I surprised the expedition, and its members were interned as enemy foreigners.

The British expedition under Arthur Eddington five years later had much better luck. It was a lesson to Einstein's scientific critics but also, in a sense, a humiliation for Germany's science policy makers. German politicians could not be oblivious to the fact that the theory of a native member of Germany's most renowned scientific institution had been confirmed by researchers from one of the victorious enemy nations. The policy being followed at that time was salvaging what prestige had so loosely been gambled away during the war by relying on German science. Right after the war had ended, Max Planck had pointed out:

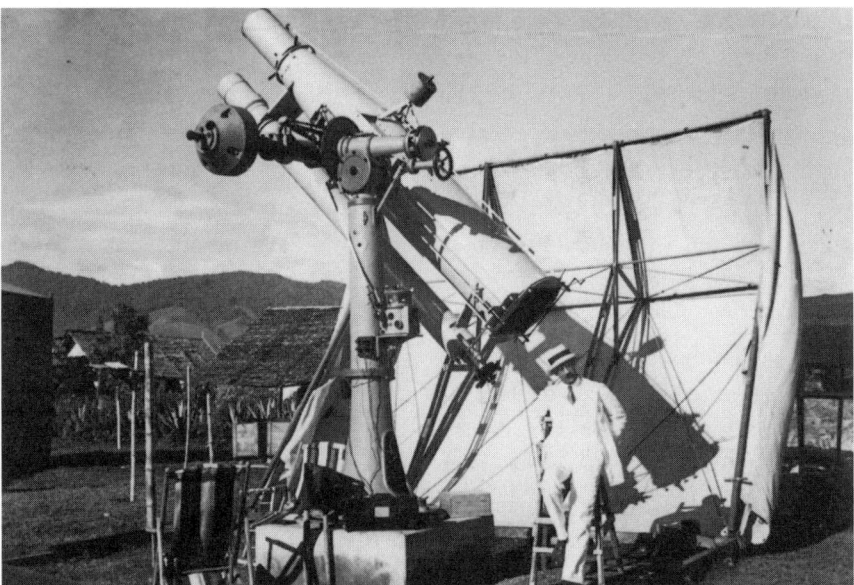

Fig. 2.27. Erwin Finlay-Freundlich during the solar eclipse expedition of 1929 in Sumatra

If the enemies of our Fatherland have taken away its defense and power, if severe crises have broken in and heavy ones are perhaps still before us, there is one thing no external or internal enemy has yet taken away from us: the position that German science fills in the world. But guarding this position and, if necessary, defending it by every means, this our Academy, as the most eminent scientific agent of the state, is also first and foremost called upon to do. (90, p. 993)

So it was no coincidence that a nonpartisan group of representatives in the budgetary committee of the Prussian Constituent Assembly filed a submission in the same year, in November 1919, "to petition the State Government, in agreement with the Reich Government, to release the necessary funds for Germany's continued collaboration with other nations toward developing Albert Einstein's fundamental discoveries and toward furthering his own research" (21, vol. 1, p. 176; 47, trans. p. 75).

This submission appears to have been approved within a matter of days along with an allocation of 150,000 marks. In view of the difficult situation that the German economy was in at that time, this was not a small sum. But it was certainly not enough to finance a meaningful research project. In a letter to the Prussian minister of culture, Konrad Haenisch, at the beginning of December, Einstein pointed out these problems:

A day or so ago I received a report by the State Budgetary Committee of the Prussian Constituent Assembly, according to which it is planned that 150,000 marks from the State Treasury be placed at my disposal in support of research in the area of the general theory of relativity. However great my joy and feeling of gratitude for this truly generous gesture, I nonetheless cannot withhold painful reservations. In these times of extreme need, would such a resolution not justifiably trigger bitter feelings among the public? I believe that we can promote research effectively in the area of the general theory of relativity even without use of special funds, if the country's observatories and astronomers would simply place a portion of their apparatus and labor at the service of this cause. Hitherto Dr. E. Freundlich at the Astrophysical Institute in Potsdam was the only German astronomer (aside from Schwarzschild) to advance the field. It would be of great service to the cause if . . . this astronomer were to receive an observer position very soon at the Potsdam institute with the objective of working on testing the general theory of relativity. (21, vol. 1, p. 176; 1, vol. 9, trans. p. 165)

In parallel with this initiative, and surely not without consulting Einstein, Freundlich, whose own scientific career was at stake, became the champion of efforts to organize the development of a large instrument that would allow another prediction of the general theory of relativity to be experimentally tested: a shift of spectral lines toward the red end of the spectrum when subjected to a strong gravitational field. Initial trials he had conducted during the war as well as those by the former director of the Astrophysical Observatory in Potsdam, Karl Schwarzschild, who had since died in 1916, had not yielded any convincing results. It instead became clear that fundamentally new technology of the highest possible precision was needed to detect the effect. In the United States the first so-called tower telescopes, built specifically for solar research, had been in use for astrophysical investigations since the turn of that century. Freundlich wanted Germany to have such a tower telescope as well, in order to use it to verify experimentally the existence of the predicted gravitational redshift. Capitalizing on the great public interest in Einstein and his theory of relativity and the state grant of 150,000 marks, Freundlich drafted an "Appeal for the Albert Einstein Donation Fund" in December 1919. The support of leading Berlin scientists was procured before sending it on to powerful funders from industry and business. In this appeal Freundlich ably linked a description of the scientific importance of the planned enterprise with an appeal to the recipients' sense of patriotism:

> Albert Einstein's work on the general theory of relativity is an important turning point in the development of science. . . . Experimental verification of its observable consequences, to prove the applicability of the new theory, must go hand in hand with further elaboration of the theory. Only astronomy seems suited to take up this task at the present time. . . . The academies in England, America, and France have recently set up a commission, excluding Germany, to busily lay the experimental foundations of the general theory of relativity. Persons concerned about Germany's cultural standing are honor-bound to provide what funds they can to enable at least one German observatory to work on the theory in direct collaboration with its author. These funds are intended for procuring the observational equipment necessary for the Astrophysical Observatory, which is placing itself at the service of this cause, to work successfully on this problem. (21, vol. 1, p. 177; 47, trans. pp. 79)

Within a year the appeal yielded a total capital of over 300,000 marks. Aside from grants of money there were also offers by industrial firms to donate the necessary equipment. For instance, two firms in Jena, Carl Zeiss and Schott & Associates, delivered the instrumental equipment at cost. The galloping inflation in Germany dictated speed in implementing the project, so the construction of the tower telescope began in the summer of 1920, before the funding had even been finally secured. It was fortunate that Freundlich already knew the architect Erich Mendelsohn for many years. This private connection took a professional turn during World War I when Freundlich contacted his acquaintance, who was stationed on the Russian front, about "new opportunities" that had opened up for him in the form of a directorship and the building of a new observatory:

> I am contemplating a project of building a little institute for my own research, once the influential individuals at the Kaiser Wilhelm Institute have approved my plans and I have been explicitly requested to submit a memorandum with blueprints. ... The director of the Potsdam Observatory has offered me a very suitable plot of land on his observatory's premises as a building site. If it can be arranged, I will try to have you prepare the blueprints for the exterior design, although it will not be a very rewarding job for you. (49, trans. p. 59)

This inspired Mendelsohn to make a few sketches that were then solidified in further dialogue with Freundlich. These early sketches resemble quite closely the final structure that materialized in 1920.

The concrete planning stage for the construction took place during the summer months of 1920, and Freundlich was able to obtain the necessary approvals for the envisioned modernist design from the board of trustees of the Einstein Donation Fund and even from the responsible government agencies with astonishingly little resistance. The elongated lower story of the tower structure evokes associations with a submarine or wind-swept dunes.

The board had practically granted Freundlich a general power of attorney for the construction work, and the runaway inflation surely also did its part. There was no time for discussion if jeopardizing successful completion of the entire project was to be avoided. According to the architect's original conceptions, its fluid form was to be constructed of steel-reinforced concrete. But neither time nor the available building technology allowed this method to be used throughout. So the building was completed as a mixed construction. The pedestal and the tower shaft were made of masonry, and the shaped parts of the main body, such as the interior stairway and the end of the tower, were made of molded concrete. The rotatable dome at the top of the tower was a wooden construction. To lend a uniform, sculpted appearance to the tower, the entire structure was finished off with a plastering of molded cement. This mixture of building techniques makes the building very susceptible to weathering, and it still requires major renovations at regular intervals.

In August 1921 the rough structure was ready for the official building inspections. Only the removal of the scaffolding and a final layering of ocher paint remained to be done before the tower could be presented to the professional world. This took place at the twenty-first convention of German astronomers, which happened to take place that year on Telegraph Hill in Potsdam. It was on this occasion that the only photograph of Einstein with "his tower" was taken.

The installation of the telescope optics and other instrumentation took more time, until the end of 1924, so it was not possible to make the first observations until 1925.

Known from the very start by its popular name, the Einstein Tower, the tower telescope is technically modeled after an instrument at Mount Wilson Observatory in the mountains near Pasadena, California. But architecturally they could scarcely be more different. In contrast to the striking design of the Einstein Tower in Potsdam, the California tower is a simple frame of naked steel. The laboratory for the spectral analysis in Potsdam is not located directly under the tower either. The light is redirected by a mirror at the foot of the tower to follow a horizontal path into a cellar area toward the rear. Being underground, the laboratory is better shielded from temperature and humidity fluctuations as well as from other disturbances. In this almost twenty-meter-long room, a prism or grating spectrograph diffracts the beam of light, and the resulting

Fig. 2.28. Albert Einstein at the entrance of the Einstein Tower, 1921

spectrum is recorded onto a photographic plate. The plan was to analyze the spectra from sunlight in daytime and from the light of brighter stars at night. It was hoped that a comparison of the spectra obtained against ones made from terrestrial light sources would reveal a gravitational redshift in solar and stellar spectral lines, thus confirming Einstein's general theory of relativity on yet another point.

These hopes were in vain. None of the spectral-line measurements performed in the astrophysical research institute, which had been named the "Einstein Institute," yielded the anticipated effect of gravitational redshift under the supervision of Freundlich. The reason for the failure was that in the 1920s the precision necessary for detecting such a subtle effect of the general theory of relativity was not yet feasible; that only became possible more than three decades later with the discovery and application of the Mössbauer effect in the relevant measurements. The investigations conducted by Freundlich and his coworkers at the Einstein Institute earned notice, nonetheless. Many new insights were made about the physics of the sun and other astrophysical matters in general as well.

The emergence of the National Socialists in government also had grave consequences for the Einstein Institute. Renamed the "Institute for Solar Physics," it was incorporated into the Astrophysical Observatory, thus permanently losing its administrative autonomy and independent endowment. Moreover, it had to change its research profile, and all work regarding the verification of relativistic effects had to cease. Freundlich lost his position and had to emigrate to Turkey in the fall of 1933, then later to Prague, and finally to Scotland. The director of the Astrophysical Observatory, Hans Ludendorff, ordered that the bust of Einstein be removed from the entrance immediately; in its place appeared a single symbolic stone: *Ein Stein*. The bronze bust, made by the Dane Harald Isenstein in 1926, survived the Third Reich unscathed in the safety of the institute's basement.

Einstein dutifully performed his obligations as chairman of the board of trustees of the Einstein Institute until 1933—he had been appointed "for life"—and therefore regularly climbed up Telegraph Hill in Potsdam, but he became less and less interested in the research being conducted there. One reason may have been that quite soon after the observational work was commenced, it became evident that the tower would probably not fulfill its original purpose. Moreover, a rift had developed in the friendship between Einstein and Freundlich in 1921 when Freundlich tried to sell a manuscript of Einstein's—to raise money for research, Freundlich maintained. From then on, all further contact between the two men was marked by polite but extreme reserve. To Freundlich's archenemy and rival Ludendorff, Einstein unequivocally wrote in 1925:

Fig. 2.29. Harald Isenstein's bronze bust of Einstein, with one stone (*Ein Stein*)

With regard to Mr. Freundlich, . . . I have broken off personal relations with him and could have added a few very fine "specimens" to the list of sins you reported. . . . But I respect his organizational achievements and act accordingly. . . . Thus we both serve the cause, even though we value the man and the scientist little. He is not worth getting upset about. (21, vol. 1, p. 196; 49, trans. p. 102)

Although the Einstein Tower did not satisfy expectations and never produced an experimental verification of the gravitational redshift predicted by Einstein's general theory of relativity, the research conducted there was very important for modern solar physics. To this day the Einstein Tower remains an internationally acknowledged center for solar physics observations. Its contributions to a better understanding of sunspots and their magnetic fields were significant. But the tower is even more noteworthy for its modern architecture. The Einstein Tower's bold and unusual shape makes it a recognizable structure throughout the world. It is the very emblem of modern expressionist architecture for architects and the general public.

CHAPTER THREE

Homo Politicus

The man who arrived in Berlin in the spring of 1914 was not yet the politically active scientist Albert Einstein was to become by the end of his career in Berlin, his iconic stance earning him fame that lasted to the end of his days. We have no record of any political statements by him predating 1914; his later so dominant political and social engagement then still lay fallow (72c). This may have been because the focus of the foregoing years had been on his "miraculous year 1905" and further developments in his pioneering science. But even in Berlin, Einstein was not yet a political activist in the fullest sense of the word. He never issued any political declarations or resolutions on his own account; he did not even belong to any particular political party. Any statements of that nature were less a product of his own political action than of campaigns and initiatives by others. At first, political views and convictions governed such reactions less than deep-rooted antimilitarism and an instinctive or moral humanism, and such comments were often only made in private. The evolution of the humanist and moralist Einstein into a *homo politicus* has much to do with the fall of the German Kaiserreich and the establishment of the first German republic. The spectacular verification of his general theory of relativity transformed him into a public figure. As a passionate democrat, Einstein had much sympathy for the social and political changes promised by the November Revolution of 1918 and did not hesitate to say so in public. This personal politicalization went hand in hand with a rediscovery of his Jewishness, leading him to become a protagonist as well as a "famed bigwig and decoy-bird" for Zionism (72b; 72c). Einstein's commitment to the Weimar Republic and his support of the Zionist movement turned him on two counts into a political outsider among the German professoriate of the 1920s. It also turned him into a target for chauvinistic and anti-Semitic hate campaigns. He was endangered not just personally but also professionally. The political right decried even his scientific theories as "Jewish" or "Bolshevist." By virtue of his political views, but also of his public fame, Einstein came into contact with a wide variety of organizations that sought his competence and prominence as a speaker and sympathizer of their causes. In granting them his support Einstein was careful to maintain his independence from them. For these organizations he was a welcome but often unpredictable billboard. He was definitely a prominent purveyor of their ideas.

Site 14. New Fatherland League
German League of Human Rights
Spichernstrasse 3
10777 Berlin (Wilmersdorf)

subway stop Spichernstrasse (U1, U9),
from there 2 min. by foot

When the First World War broke out in August 1914 and Germany was gripped by a wave of chauvinism and nationalism, Einstein felt strongly repulsed by such fanaticism. Depressed about it, he wrote to his friend and colleague in Leyden, Paul Ehrenfest: "At such times one sees to what a deplorable breed of brutes we belong [...] and [I] feel only a mixture of pity and disgust" (1, vol. 8, trans. p. 41).

Thus he would have no part of all the patriotic cheering and agitation that many of his colleagues unreservedly and enthusiastically applauded. His signature does not appear under the infamous "Appeal to the Civilized World"—its initiators surely did not even consider him a potential cosigner. With this appeal Germany's intellectual elite legitimized German militarism under the aegis of German culture: "Without German militarism, German culture would have been obliterated from the face of the earth long ago. The German army and the German nation are one" (93, p. 25).

Such pithy words were espoused by those who had arranged for Einstein's transferal to Berlin; they numbered among his nearest colleagues and acquaintances: the physicists Max Planck, Walther Nernst, and Fritz Haber were among them as well as left-wing intellectuals Gerhart Hauptmann and Max Reinhardt.

It is indicative of Einstein's independence, not just on scientific questions, that he would go against the political mainstream, to which his famous and personally respected colleagues belonged, when he put his signature under the

Fig. 3.1. The "splendid banquet halls of the west" of the city, at Spichernstrasse no. 3, were destroyed in World War II.

counterappeal "To the Europeans!" drafted by the Berlin physiologist and pacifist Georg Friedrich Nicolai. This manifesto advocated the power of reasoned politics, a rapid end to the war, and general understanding among nations (1, vol. 6, pp. 69–70; 47, trans. pp. 29–30). Although the existence of this manifesto did not become generally known and it was only published after the war, it did not go unnoticed. Other opponents of the war became interested in Einstein. Probably through Nicolai, and perhaps also through Count Georg von Arco, the engineer and pioneer of wireless telegraphy, Einstein made the acquaintance of activists of the Bund "Neues Vaterland." This New Fatherland League had been founded in Berlin in November 1914, and its administrative office was located initially at Tauentzienstrasse no. 9, later at Kurfürstenstrasse no. 126, close to the zoo, and then at Wilhelmstrasse no. 48 in the government quarter of Berlin, before the association moved into its own building at Monbijouplatz no. 10 in the second half of the 1920s.

The league quickly distinguished itself as an organization of left-wing intellectuals whose aim was a rapid end to the war and the establishment of a postwar order based on peaceful competition among the European states. Besides this pacifistic orientation, their domestic positions favored the introduction of universal suffrage, equality between men and women, and parliamentary democracy. The league publicized its goals in public notices and other publications as well as by organizing public events in downtown venues such as the former Prussian Herrenhaus (Leipziger Strasse no. 3), the Demokratischer Club (Viktoriastrasse no. 24), the Spichern banquet halls, or the parliament building, the Reichstag.

One myth perpetuated in some publications about Einstein is that he was a founding member of the league. The source of this error can be traced back to one of the cofounders, the journalist Otto Lehmann-Russbüldt. In his book about the history of the league, he counts Einstein among its ten "first members and sympathizers." Einstein's membership card actually bore the number 29, and he joined only in March 1915. This distortion of the historical facts was surely not without practical purpose. When the book was published in 1926, the publicity magnet Einstein was certainly an effective form of promotion for the successor organization to the league, the League of Human Rights. Einstein's name is first mentioned in the association's records on March 21, 1915, when its fifth meeting convened in the "Haus des Deutschen Sports" on Schiffbauerdamm. Einstein is listed there together with his future wife, Elsa, under the rubric "guests." Another guest at this meeting was the expert on international law and pacifist Walther Schücking, who noted about his encounter with Einstein:

> Professor Einstein was also there; it was the first time I had heard of his name. Through a law he had discovered regarding the universality of time he is supposed to have accomplished a scientific feat of the very first order and has for that reason been drawn away from his Swiss home country by the Ministry of Culture in order to devote himself to research in Berlin with no teaching obligations. (47, trans. p. 33)

Right after joining the league, Einstein took part in one of its initiatives. Together with Count Arco and the writers Rudolf Goldscheid and Leo Kestenberg he was supposed to compose an "appeal by intellectuals" for peace and international understanding. However, it is not clear whether a final draft was ever drawn up—if one was, it remained a private matter among individuals. Einstein's contact with the French writer and pacifist Romain Rolland also remained a more or less private matter, although the league likewise played an important part in mediating their acquaintance. On March 22, 1915, one day after their first meeting, Einstein wrote to Rolland:

> From the daily paper and through my connections with the highly creditable "Fatherland" League I learned of how courageously you placed your life and person at risk toward eliminating the so ominous misunderstandings between the French and German peoples. . . . I place my feeble powers at your disposal in case you think I can be of service to you as a tool, be it through my residence or through my connections with German and foreign individuals in the exact sciences. (1, vol. 8, trans. p. 77)

Rolland immediately took up Einstein's offer of assistance and asked him for support in organizing an inspection by the International Red Cross of prisoner-of-war camps in Germany. Traces of any concrete action Einstein might have taken are lost to posterity—here as elsewhere it appears that Einstein's good intentions or verbal commitments were greater than his actual deeds. In any event, a personal meeting did take place between the two. During one of Einstein's trips to Switzerland to see his children in September 1915 he visited Rolland, who at that time was living in Vevey by Lake Geneva. About this encounter Rolland noted in his diary:

> Professor A. Einstein, the ingenious physicist and mathematician at the University of Berlin, who has written me during the course of the past winter, comes to visit me from Zurich. . . . Einstein is unbelievably frank about his opinion of Germany, where he lives. No German has such frankness. Any other person would have suffered from the sense of isolation as a thinker during this terrible year. Not he. He laughs. He has managed to write his most important scientific work during the war. I ask him whether he mentions his views in the presence of his German friends and discusses them with them. He says no. He contents himself with posing many questions—as Socrates did—to upset their peace of mind. He adds, "those people don't like it very much." (91, p. 697; 47, trans. p. 39)

"Those people" were not the only ones to dislike such questions. The German authorities liked them as little as they appreciated the pacifist activities of Einstein and the other members of the league. Through arrests, censorship, and other repressive measures the league's activities were progressively reduced until it was banned outright on February 7, 1916. Einstein—along with other members of the league—had already become the object of surveillance by the police. The postal censors had first intercepted Einstein's postcard to the Dutch peace association, the Anti-Orloog-Rad, in December 1915, which led to an inquiry by the Berlin police. The resulting report stated that Einstein was a member of the New Fatherland League "but has not yet drawn any attention to himself through any agitatory behavior within the pacifist movement. From a moral point of view he enjoys the best reputation conceivable and there are no penal records on him. He has a subscription to the paper *Berliner Tageblatt*" (23, p. 309; 47, trans. p. 35).

After the league had been dissolved, the authorities continued to keep a close eye on Einstein and other peace activists. They complained about him to the academy: "Dr. Einstein has left repeatedly on travels without personally registering his departure here in Berlin at the Police and without personally registering his arrival with the Police at his travel destinations, which he as a neutral foreigner is obligated to do" (21, vol. 1, p. 198; 47, trans. p. 36).

There is no doubt about Einstein's pacifist convictions, and his statements about the political situation were unequivocal. But during the First World War Einstein was anything but a political activist. After the banning of the league, Einstein withdrew quite completely from the activities of the peace movement, and his political remarks were limited to the private sphere again—they appear in his letters to his friends and colleagues like Paul Ehrenfest or his encounters with Romain Rolland or other contemporaries. The absence of any concrete political convictions behind his pacifist activities is also evidenced by Lise Meitner's report about a musical evening at the Plancks' home in the fall of 1916. Einstein performed there not just as solo violinist; he was also "emitting all the while such pricelessly naive and peculiarly political and warlike quips. The mere fact that such a cultivated person exists who in these times never ever picks up a newspaper, is certainly a curiosity" (96, L. Meitner to O. Hahn, Nov. 16, 1916).

Einstein's public restraint and political naïveté changed only under the influence of the revolutionary upheavals of November 1918. Then Einstein again publically aired his views about his pacifistic, democratic, and socialistic ideals, and his political views became increasingly concise.

While the November Revolution was under way, he gave a talk at a meeting of the newly founded league in the Spichern banquet hall at Spichernstrasse no. 3, in which he not only announced his support of the new government led by the Social Democrats but also formulated his political convictions and hopes:

> Fellow members, ladies and gentlemen! Allow an old democrat, who has not had to readjust politically, a few words. Democracy marks our times, i.e., the reign of the people. It is only feasible if the individual holds two things sacred, namely, a belief in the sound judgment and sound will of the people. A willing subordination to the

Fig. 3.2. One of the Spichern banquet halls, around 1930

will of the people pronounced by vote and election, even if this will of the people is contrary to one's personal will or ideal. (23, p. 327)

In the following years Einstein's membership in the New Fatherland League or its successor organization, the League of Human Rights, was distinguished by brief service on the association's governing board. He was reelected to the board for the term year 1925–26, and from 1928 until his emigration he was a member of its political council, which offered its advice and expertise on the board's political activities.

By assigning Einstein such offices the association was using him as their billboard, of course, but these representative functions were only one element of his involvement. He also actively participated in a number of its activities and events. On December 16, 1919, for instance, during a demonstration organized by the league, he delivered the welcoming speech to Paul Colin, editor of the periodical *L'Art libre* from Brussels, as "the first Frenchman to come and see us after the war in the service of the sacred goal of mutual understanding among nations" (8, p. 56).

In February 1921 he even traveled as the league's emissary to Amsterdam along with its managing director Otto Lehmann-Russbüldt and Count Kessler to discuss the repressive reparation demands made by the Allies with the newly founded International Confederation of Unions. Their purpose was to ask the confederation to ply its influence on the major powers in this matter. As on other occasions, Einstein's particular negotiation skills were not called for. He simply accompanied the two politicians as a "famed bigwig and decoy bird"—as he once ironically described himself in another context—in order to lend weight to the issue by virtue of his public prominence and international acclaim.

When the New Fatherland League was renamed the German League of Human Rights in 1922, Einstein continued to participate in its activities. His engagement in the league was defined by his personal commitment to mutual understanding between the French and Germans. He was one of the signers of the appeal "To Democrats of Germany and France," launched in February 1922 by the league with its sister association in France, demanding general disarmament as well as the resumption of scientific and cultural relations between the two countries. His spectacular trip to Paris in spring 1922 was likewise directly connected with the league's international activities. In the mid-1920s, when international relations were beginning to normalize among scholars, the focus of Einstein's sociopolitical activities shifted back to his pacifist persuasions, and his message concerned general disarmament and the protection of human rights. He carried this message beyond his involvement with the league with numerous private initiatives and appeals. Einstein's pacifist stance became more sharply defined. He publicly supported antiwar conscientious objectors and lambasted the military as a "disgrace to civilization." In a letter to Frieda Perlin he unequivocally stated in 1928: "To me any killing of human beings is common murder, even if it is the state doing so on a large scale" (23, p. 440).

Einstein's last major effort for the league probably took place in spring 1931. He joined the initiative "against the university reactionaries" in support of the mathematician and civil rights activist Emil Julius Gumbel, who had become the brunt of aggressive rightist attacks and ostracism in the second half of the 1920s because of his activities as a pacifist and socialist. The University of Heidelberg even threatened to revoke his permission to teach. Einstein repeatedly stood up for Gumbel, stating: "The conduct of the academic youth against him is one of the most tragic signs of the times, esteeming so little the ideals of justice, tolerance, and truth. What will become of a nation that persecutes such contemporaries and whose leaders pose no resistance to the mean masses?" (53, p. 109; 47, trans. p. 255).

When another defamation campaign was launched against Gumbel, the league organized an event to demonstrate its solidarity with him on April 27, 1931. The venue was Langenbeck-Virchowhaus at Luisenstrasse no. 62. (After

Fig. 3.3. An event of the League of Human Rights, Berlin, 1932. The historian Martin Hobohm is speaking at the rostrum; seated *(from left to right)* are the historian Arthur Rosenberg, Albert Einstein, and the mathematician Julius Gumbel.

World War II the People's Chamber of the East German Republic convened there and, beginning in 1976, the Academy of the Arts; today it serves as a venue for medical-science gatherings.) Einstein had been one of the organizers of this event, which drew a packed house of more than a thousand people. He rose briefly to speak "about the freedom of teaching":

> Among the row of speakers, Einstein was saved up for just before Gumbel's closing speech. The tension grew enormously: What will he say? When the welcoming applause finally died down, Einstein stated that to him the topic of that evening seemed to have already been sufficiently covered but one thing he did want to say: one shouldn't just cheer at Gumbel but should also read his writings more than before, including his book on political murders. "I myself have learned many things from this book and I believe you, too, can learn something from it."—Thus he spoke and was gone. The vague unease in the mighty, overcrowded hall was palpable; the obligatory, overwhelming applause salvaged the situation. But Gumbel's book could hardly have been more effectively impressed upon their minds.
> (53, p. 122)

Another equally impressive closure to Einstein's active involvement with the New Fatherland League and the League of Human Rights, was a gramophone recording arranged by the league made in the fall of 1932 in Berlin. Under the heading "My confession of faith" Einstein revealed something about his view of the world and politics, stating among other things:

> My passion for social justice has often placed me at odds with people, likewise my dislike of any kind of bondage and dependence that did not appear to be absolutely necessary to me. I always respect the individual and have an irrepressible abhorrence of force and the exclusive club mentality. From all these motivations I am an impassioned pacifist and antimilitarist, rejecting all forms of nationalism, even if it just poses as patriotism. (28, p. 100)

The gramophone record, which must have appeared around the beginning of 1933 did not find much circulation, however. Neither the content nor the author were the cause. In 1933 the political situation had changed. Neither the league nor Einstein was amenable to the Nazis. Their early and consistent resistance to National Socialism had turned them early on into objects of ostracism, repression, and persecution, forcing Einstein into exile and the association into prompt self-imposed dissolution. Einstein himself surely would only have got hold of the record in exile, because he left Berlin for his lectureship in America just a few weeks after the recording session, never to return to Germany.

Site 15. Parliament building
Platz der Republik 1
11011 Berlin (center)

subway and met. train sta. Friedrichstrasse
    (U6, various S lines) or
met. train sta. Unter den Linden (S2, S25),
from there 5 or 3 min., respectively, by foot

The end of the First World War led to the November Revolution and the collapse of the German kaiserreich. Albert Einstein was euphoric. "The great event has taken place! . . . That I could live to see this!! No bankruptcy is too great not to be gladly risked for such magnificent compensation. Where we are, militarism and the privy-councillor stupor has been thoroughly obliterated" (1, vol. 8, trans. p. 693), he wrote to his sister in those revolutionary days of November, and he reassured his mother on the same day: "Don't worry. All has been going smoothly, impressively even, up to now. The current leadership really seems to be equal to the task. I am very pleased with the matter's development. Now I'm beginning to feel really comfortable here. The bankruptcy has done wonders" (1, vol. 8, trans. p. 693).

Fig. 3.4. The parliament building (Reichstag) at the time of the November Revolution, 1918

Coming from a German academic, such a profession of faith in the November Revolution and the political changes taking place in Germany at the end of World War I was extremely rare. The majority of the professoriate mourned the demise of the kaiser's empire, his value system and authoritarian maxims, and not least of all, his power and influence.

Thanks to this reputation—as Einstein also wrote his friend Michele Besso—"of an irreproachable Socialist . . . , yesterday's heroes are coming fawningly to me in the opinion that I could break their fall into emptiness. Funny world!" (1, vol. 8, trans. p. 703).

So it was certainly no coincidence that during those tumultuous days of the Räterepublik, Einstein was asked by his university colleagues to mediate with the students' council, which had established itself at the University of Berlin like the workers' and soldiers' councils everywhere else in Germany had established themselves and were brandishing their political power. One of its first measures had been to close down the university and detain the conservative rector Reinhold Seeberg and other university dignitaries. On the hope that the *"Obersozi"* (high-placed Red) Einstein would be accepted by the radical students and perhaps be able to exert a moderating influence on them, he was asked to start negotiating with the students' council and, to begin with, obtain the release of their detained colleagues. Einstein accepted the proposal and joined the delegation of negotiators, which, in addition to Einstein, consisted of his friend and fellow physicist Max Born and the psychologist Max Wertheimer. Max Born gave a retrospective report about their mission:

> I will not go into the difficulties we had in penetrating the dense crowds which surrounded the Reichstag building and the cordon of revolutionary soldiers, heavily armed and red-beribboned. Eventually someone recognised Einstein, and all doors were opened.
>
> Once in the Reichstag building, we were escorted to a conference room where the student council was in session. The Chairman greeted us politely, and asked us to sit down and wait until an important point in the new statutes for the university had been dealt with. So we patiently waited and listened. Eventually the point at issue was settled and the Chairman said: "Before we come to your request, Professor Einstein, may I be permitted to ask what you think of the new regulations for the students?" Einstein thought for several minutes, and then said something like this: "I have always thought that the German universities' most valuable institution is academic freedom, whereby the lecturers are in no way told what to teach, and the students are able to choose which lectures to attend, without much super-

vision and control. Your new statutes seem to abolish all this and to replace it by precise regulations. I would be very sorry if the old freedom were to come to an end." Whereupon the high-and-mighty young gentleman sat in perplexed silence. Then our business was discussed; but the student council decided that it had no authority in the matter, and referred us to the new Government in the Wilhelmstrasse, issuing us with a pass for this purpose.

Accordingly we walked on to the Reich Chancellor's palace. This was a hive of activity. The footmen of the Emperor's time still stood in corners of the passageways and stairs but, apart from them, the people running about the corridors were more or less shabbily dressed and carrying briefcases—socialist delegates and delegations from the workers' and soldiers' councils. The main hall was full of excited people talking in loud voices. But Einstein was recognised at once, and we had no difficulty in getting through to the newly appointed President Ebert, who received us in a small room and said that we would appreciate that he was unable to pay attention to minor matters that day, when the very existence of the Reich itself was in the balance. He wrote a few words on our behalf to the appropriate new minister, and in no time at all our business had been concluded.

We left the Chancellor's palace in high spirits, feeling that we had taken part in a historical event and hoping to have seen the last of Prussian arrogance, the Junkers, and the reign of the aristocracy, of cliques of civil servants and of the military, now that German democracy had won. Even the long journey back to the Grunewald, mostly on foot, could not dampen my elated mood. (16, trans. pp. 150–151)

A quarter of a century later Einstein's and Born's exultation had completely collapsed along with their hopes for political rationality. Both had been driven away from Germany by the Nazis, were living in exile and were thinking about what should happen in Germany after the anticipated defeat of Hitler's Germany. Einstein wrote resignedly to his friend in this regard:

Do you still remember the occasion some twenty-five years ago when we went together by tram to the Reichstag building, convinced that we could effectively help to turn the people there into honest democrats? How naive we were, for all our forty years. I have to laugh when I think of it. We neither of us realised that the spinal cord plays a far more important role than the brain itself, and how much stronger its hold is. I have to recall this now, to prevent me from repeating the tragic mistakes of those days. (16, trans. p. 148)

Fig. 3.5. Max Born, around 1920

Site 16. Old Philharmonic Hall
Bernburger Strasse nos. 21/22a
10963 Berlin (Kreuzberg)

subway and met. train sta. Potsdamer Platz
    (U2, S2, S25) or
met. train sta. Anhalter Bahnhof (S2, S25),
from there 5 or 3 min., respectively, by foot

The spectacular verification of the general theory of relativity in 1919 by a British solar eclipse expedition transformed Einstein into a public figure, particularly since his personality and his accomplishments were colored with qualities that actually had nothing to do with his scientific work. Einstein's name was no longer the domain of solely physics and science. He was looked upon not just as a genius physicist but also as a democrat, a pacifist, and a Jew. The controversy surrounding Einstein and his theory at the beginning of the 1920s was thus no longer among scientists alone. It began increasingly to take place on the public stage, with the opponents of relativity theory ideologically objecting to its unintuitiveness and the complicated mathematics it required. According to their position, anything that is not simple and intuitive to "healthy common sense" could not be true. Moreover, parallels were being drawn at that time between the disintegration of the political order, the relativization of social values hitherto regarded as unshakable, and the revolutionary changes going on in science and specifically in physics. Antimodern prejudices mixed with anti-Semitic and nationalistic tendencies led the political right in Germany to regard relativity theory as "Bolshevist" and "Jewish" and to make Einstein himself the target of malicious and predominantly racist attacks.

Fig. 3.6. The main hall of the old Philharmonic, around 1930. The building was destroyed in World War II.

These attacks reached their first high point in the summer of 1920, when the "Syndicate of German Scientists for the Preservation of Pure Science, reg. assoc." announced a series of about twenty lectures against relativity theory. The initiator of this drive was Paul Weyland, who up to that point had made a name for himself only as the author of shallow rightist texts. Thinking he had now deciphered the signs of the times, he felt compelled to express his and others' discontent about the current social developments with a campaign against Einstein's theory of relativity. The goal of his campaign was—as we read in the clumsy German of a flyer—to "protect the German people from being misled by overly highly acclaimed scientists (in some circles) who set the scientifically interested world in disarray with half-baked opinions" (45, p. 180).

For this purpose he tried to secure the support of prominent scientists known for their criticism of Einstein's theory, and some of these, whom Weyland considered appropriate critics, discovered their names on the list of scheduled speakers without even first being informed about it. This led to numerous cancellations. In the end, only two of the scheduled twenty events actually took place, and only the first, on August 24, 1920, in the large hall of the Berlin Philharmonic, attracted public attention.

On that occasion Weyland was the first to speak on the subject "observations on Einstein's theory of relativity and the manner of its introduction" (23, pp. 337–348). As can be gathered from a report in the *Vossische Zeitung*, he ran out "the heavy artillery":

> He countered the "Einsteinian fictions" without devoting a single word to explaining what they were actually about. Physicists who took Einstein's part were strongly suspected, he personally and his friends, accused of having set up the daily papers and even professional journals for the purpose of publicizing the theory of relativity. Having still not learned what it was actually about, the public began repeatedly to call out: "Let's get to the point!" Mr. Paul Weyland retorted to this friendly suggestion: "Appropriate measures have been taken to deposit scandal mongers out of doors!" Further invectives against the clique of professors, for which the speaker borrowed heavily from Schopenhauer, were followed by lamentations about the lowering intellectual niveau of our nation. (50, p. 57)

At the close of his speech Weyland lambasted relativity theory as the product of an intellectually confused time and called it "scientific Dadaism." The cat was out of the bag. This was no harmless rhetoric. By linking relativity theory to Dadaism, Weyland was consciously insinuating a diabolical intent of the theory to corrupt physics into a degenerate science—as degenerate as the supposed meaningless stammerings of Dada poetry and prose, offensive to healthy common sense and ultimately also offensive to sound national sensitivities. Twelve years later such views were no longer those of only outsiders and troublemakers but had become a component part of National Socialism.

Another speaker on the program that evening was the physicist Ernst Gehrcke, who in contrast to the *spiritus rector* of the event was neither a scientific charlatan nor an anti-Semitic demagogue. He was a serious physicist who had earned his standing with important research on optics, making a name for himself as a leading experimental physicist. As such he observed with great dismay the rising importance of theoretical physics and the revolution simultaneously taking place in physics. He could only regard the new and far too speculative-looking theories of modern physics with skepticism—especially when their experimental verification was proving problematic. He was following the ideal of the classical scientist, who—in Gehrcke's own words—sought his "sphere of activity in the area of exact trials and their logical interpretation" and encountered "phantastical theories and boundless speculations, whose findings are draped in a mathematical dress that often stands reality on its head" with reservations, if not

Fig. 3.7. Paul Weyland, 1920s

downright opposition (56, p. 75). Gehrcke regarded Einstein's theory of relativity as a model example of a "phantastical" theory whose boundless speculations and highly developed mathematical formalisms led to results far from reality. However, Gehrcke's conceptions did not receive much approval in the physics world, because the overwhelming majority of his contemporaries regarded the special theory of relativity as a secure pillar of knowledge in their field. Gehrcke and his fellow antirelativists were thus becoming increasingly sidelined, particularly as Einstein's popularity grew at the beginning of the 1920s. This prompted Gehrcke and other opponents of relativity theory to conclude that no attention was being paid to their years of scientific criticism because the relativists were resorting to exaggerated publicity and mass suggestion. As evidence Gehrcke collected more than five thousand newspaper articles related to Einstein and relativity theory. Thus Gehrcke embarked on a new phase of his opposition to relativity theory. He gradually moved away from the range of scientific discussion and publications into public polemics and occasionally even into political disputes. This tendency made him an obvious choice for a speaker at the Berlin Philharmonic and, at least temporarily, an ally of Paul Weyland.

The difference between these two opponents of Einstein could hardly have been greater though, as even the audience at the event began to realize. Max von Laue reported in a letter to his Munich colleague Arnold Sommerfeld that although in his talk Gehrcke "raked up the same old stories, his calm, factual manner of speaking was a relief after Weyland, who can measure up to the most unconscionable of demagogues" (92, vol. 2, p. 80). A newspaper report remarked that Gehrcke "was visibly disconcerted after this apparently unexpected introduction. But his voice soon became steadier again and he presented his reservations about the theory of relativity in a soothingly calm manner" (56, p. 76).

Despite the evidently factual and unpolemical way that Gehrcke presented his antirelativistic arguments at the Berlin Philharmonic, his appearance there— just as his debate with Einstein in general—seriously damaged his scientific reputation. But it also led to a breaking of the unholy alliance with Weyland, which did not mean, of course, that he gave up his rejection of relativity theory. He continued to stay in contact with Einstein opponents inside and outside the country and industriously collected newspaper articles on the topic, which became the basis of his major antirelativistic works: *Massensuggestion der Relativitätstheorie* and *Kritik der Relativitätstheorie* (*Mass Suggestion of the Theory of Relativity* and *Critique of the Theory of Relativity*), which were both published in 1924 by a small press in Berlin.

Einstein himself attended the event at the Berlin Philharmonic in person. From one of the private boxes he and his stepdaughter listened to Weyland's hate tirades against his theory of relativity. His colleagues Max von Laue and Walther Nernst were also among the audience. Although Einstein reportedly followed the arguments of his opponents "in relaxed composure, at times even quietly smiling," this visit to the Philharmonic was anything but a pleasure for him. The vicious attacks must have deeply hurt him, because three days later the *Berliner Tageblatt* published a long essay by him. Just the day before, a public statement of solidarity by Laue, Nernst, and Rubens against the vile denigrations of Einstein and his work appeared. Under the ironic heading "My Reply—On the Anti-Relativity Theoretical Co., Ltd.," Einstein confronted his opponents and their arguments against relativity theory head on. But he also leveled very polemical attacks against the antirelativists (1, vol. 7, pp. 345–347). He specifically named not only Weyland and Gehrcke but also the Nobel laureate Philipp Lenard. The latter was the most prominent and scientifically accredited among Einstein's opponents and had not yet strayed away from solid scientific arguments. His antirelativist writings had been on sale at the Philharmonic, and his name was also listed as a future speaker on the lecture series. This probably motivated Einstein to the harsh assessment "I admire Lenard as a master of experimental physics; however, he has yet to accomplish something in theoretical physics, and his objections to the general theory of relativity are so superficial that I had not deemed it necessary until now to reply to them in detail" (1, vol. 7, p. 345).

This qualification by Einstein made Lenard into an enemy and only fanned the flames of the debate. Another remark, although by no means inappropriate, was also unfortunate. Einstein stated that he had reason to believe "that there are other motives behind this undertaking than the search for truth. (Were I a German national, whether bearing a swastika or not, rather than a Jew of liberal international bent . . .)" (1, vol. 7, p. 345). It was unfortunate because public attention was distracted from the factual content of the controversy and an anti-Semitic component was added to the discussion. That is why even well-intentioned readers and colleagues of Einstein criticized his reply as "unfortunately very clumsy" and Einstein himself soon realized that in his irritation he had overshot his target by far. "Everyone has to offer his sacrifice on the altar of stupidity from time to time, for the amusement of God and man. And I did a thorough job of it with my article" (16, p. 58; 1, vol. 10, trans. p. 265), he wrote to Max and Hedi Born.

Another remark in his correspondence with Born reveals how very deeply he had been affected by the campaign launched against him: "During the first moment of onslaught, I really did consider flight. But then I thought better of it and the old phlegm returned" (16, p. 58; 1, vol. 10, trans. p. 265).

Einstein's self-composure, or "phlegm," was surely not the only thing to persuade him to stay in Berlin. Many of his colleagues and friends sent him expressions of sympathy and solidarity. One declaration of solidarity he received was signed by the actors Alexander Moissi and Max Reinhardt as well as by the writer Stefan Zweig. In it they vented their indignation about the "pan-German incitement" and affirmed to Einstein "in truly international spirit, the sympathy of all liberal persons, who are proud to know you rank yourself among them and number among the leaders of science worldwide" (47, trans. p. 102).

The rumor that Einstein was planning to leave Berlin even reached the Prussian Ministry of Culture, leading the minister to think some reaction necessary. In an open letter to Einstein dated September 6, 1920, Konrad Haenisch wrote how he had gathered from the press

> with pangs of pain and embarrassment . . . that the theory you champion has been the object of malicious public attacks extending beyond the bounds of factual judgment and that even your professional integrity did not remain safe from disparagement and libel. It is with special gratification for me that scholars of acknowledged reputation . . . are standing by you with regard to this affair. . . . As the best are supporting your cause, it will be that much easier for you not to give any more attention to such ugly activities. Thus I may surely also express the determined hope that there is no truth to the rumor that due to these nasty attacks you wish to leave Berlin, which has always been proud, and always will remain proud, to count you, highly esteemed Professor, among the most brilliant jewels of its science. (21, vol. 1, p. 203; 1, vol. 10, trans. p. 262)

Einstein immediately replied, pointing out that "Berlin is the place in which I am most deeply rooted through personal and professional ties. I would follow a call outside of the country only in the case that external circumstances force me to do so" (21, vol. 1, p. 204; 1, vol. 10, trans. p. 263).

Although in the years that followed there were many more occasions for Einstein to consider leaving Berlin, such "external circumstances" occurred only when the National Socialists seized power in 1933. These Einstein opponents of an entirely different kind, with whom Weyland, Lenard, and other anti-Einsteinians had early on entered into an unholy alliance, eventually drove Einstein out of the city—and this time it was forever.

The Philharmonic was also the place where Einstein delivered what was probably his last public speech in Berlin. He appeared there in Beethoven Hall on

Fig. 3.8. Albert Einstein delivering a speech on October 16, 1932, in Beethoven Hall of the Berlin Philharmonic

Sunday, October 16, 1932, to speak "about the theory of relativity" as part of the program of the confederation of Jewish student associations (Verband jüdischer Studentenvereine).

Site 17. Society of Friends of the New Russia
Prussian Upper House
Now the seat of the Bundesrat
Leipziger Strasse 17
10117 Berlin (center)

subway and met. train sta. Potsdamer Platz
    (U2, S2, S25),
from there 2 min. by foot

"By the way, I must confess to you that the Bolsheviks do not seem so bad to me, however laughable their theories. It would be really interesting just to have a look at the thing at close quarters" (16, trans. p. 22), Einstein wrote to his friend Max

Born in January 1920. Despite this avowed interest, Einstein never visited the Soviet Union—too great was his fear that such a trip would excite too much propagandistic attention and be used politically in Germany as well as in the Soviet Union. However, Einstein's vibrant interest in the socialist experiment as well as his internationalism led to his taking on a leading role in the Society of Friends of the New Russia (Gesellschaft der Freunde des Neuen Russland), founded on June 27, 1923, in Berlin. It had developed out of initiatives arising from the International Workers Relief and the New Fatherland League. Not coincidentally, the founding of the society happened at a time when the Treaty of Rapallo had prepared the political ground for a normalization of relations between Germany and the Soviet Union, which had practically been brought to a standstill after the October Revolution and the establishment of Soviet power. On the flanks of the political rapprochement between the two countries, associations such as the Society of Friends of the New Russia were intended to help promote intercul-

Fig. 3.9. The Prussian Upper House (Herrenhaus) on Leipziger Strasse (now the Bundesrat of the German Federal Parliament), 1920s

tural relations. It had been founded, in a sense, as a counterpoint to the "Russian Institute," which closely cooperated with the German Society for the Study of Eastern Europe, founded as early as 1912. This institute enjoyed the strong support of the Foreign Office and other governmental agencies in its semiofficial mission to promote government policies. The conservative and often even anti-Communist attitude of its members formed the counterpart to the Society of Friends of the New Russia, whose founding, according to a letter from the Foreign Office, "came from groups with close ties to the local Soviet embassy."

The fact that a Communist functionary, the journalist Erich Baron, managed the affairs of the society was thus certainly not mere coincidence. But not all its members were covert Communists. They rather represented a broad spectrum of leftist and Russophile intellectuals recruited from the New Fatherland League. The league developed pro-Soviet initiatives very early on and particularly advocated the resumption of cultural and economic ties. Einstein's name also appears under related appeals and declarations. In September 1921 the Moscow paper *Isvestiya* published a letter by Einstein appealing to scientists in western European countries to send aid packages to their Russian colleagues to help reduce the current material shortages. Einstein wanted to do everything within his power to repair and strengthen relations between German and Russian scholars.

This open-minded attitude toward the problems of the young Soviet state also brought Einstein into contact with the Russian embassy in Berlin. He became a more or less regular visitor in the 1920s. Lydia Pasternak-Slater, a daughter of the painter Leonid Pasternak and sister of the later Nobel laureate in literature Boris Pasternak, recalled in the 1980s from the time she lived in Berlin that the embassy building on Unter den Linden was one of Berlin's cultural meeting places:

> One went over there to go to concerts, readings, short theater pieces, or for casual get-togethers . . . ; one time my mother was playing the grand piano when someone asked her if she wouldn't also accompany Einstein who was also present. But Einstein objected. He said: "I really wouldn't dare to come forward, after such an artist!" But my mother talked him into it. My father made a sketch. And that's how the fine page with the violin-playing Einstein came to be. (22, p. 489)

When the Society of Friends of the New Russia was established in June 1923, Einstein was not present at the founding meeting, but his name appears under the founding manifesto and among the members of the society's central committee—alongside the president of the Reich Parliament Paul Löbe, the archi-

Fig. 3.10. Albert Einstein playing the violin, drawing by Leonid Pasternak, around 1923

tect Hans Poelzig, and Thomas Mann. The society's purpose was to organize lectures by prominent representatives of Soviet science and politics, and its periodical *Das neue Russland* aimed to help remove the information deficit that clearly existed in Germany about scientific and cultural developments in the Soviet Union. Thus the society was not just a vehicle for the advancement of scientific cooperation between Germany and the Soviet Union but also a component of Soviet foreign policy. Its activities were carried out in close consultation with the Soviet Society for Cultural Ties Abroad, which monitored Soviet scholars' relationships with foreigners. In Berlin the society had its office at Kavalierstrasse no. 10 (now no. 22) in the Pankow district. Auditoriums and similar public spaces of educational establishments or other public institutions of the city were used for the lecture events organized by the society's individual sections.

Einstein's association with the society undoubtedly provided publicity, but he was not just a passive figurehead. He actively involved himself and made many public appearances for the society. He joined Count Georg von Arco, one of the most important protagonists of the society and chairman of the section for technology, in personally inviting a speaker to give a lecture at one of the society's events. On November 29, 1926, the Russian geochemist and vice president of the Academy of Sciences of the USSR, Alexander J. Fersman, appeared in the plenary hall of the former Prussian Upper House of Parliament (Herrenhaus) on Leipziger Strasse to speak about "successes in science and technology in Soviet Russia." According to a newspaper article about the event, Einstein also took part in the discussion and was interested in the stance taken by Russian research on international scientific cooperation. This was clearly not just a polite gesture toward the speaker, who conducted geochemical and mineralogical expeditions of international magnitude, but was also rooted in Einstein's basic understanding of the international character of science. He perceived that specifically in the Soviet Union such scientific internationalism was under threat.

Einstein and Fersman met again half a year later, when Fersman was visiting Berlin as one of eighteen Soviet scholars taking part in a Russian Scientists' Week in June 1927 to inform the German scientific community about recent

Fig. 3.11. Albert Einstein on the managing committee for a Society of Friends of the New Russia event on November 29, 1926. Count Arco is standing.

results of Soviet science. This time the Society for the Study of Eastern Europe was responsible for the event, which does not mean, of course, that Einstein would not be sighted at this occasion as well. These science weeks continued to be held every year until the National Socialists took over. They promoted international scientific cooperation while at the same functioning as Soviet cultural propaganda.

This is evidenced by the fact that the Soviet scholars were accompanied by high-ranking government representatives and party functionaries—one of them being the people's commissar for instruction and education, Anatoli Lunatsharski. Einstein first made his acquaintance at the Scientists' Week of 1927, and the two took such a liking to each other that they saw each other fairly regularly whenever Lunatsharski visited Berlin. In 1930 Lunatsharski jotted down the following observations about Einstein:

> Einstein's eyes are short-sighted and at the same time distracted. It seems as if he had long since directed half of his glances inward, once and for all. It seems as if a large portion of Einstein's power of vision and thought is constantly busy with some form of calculation. His eyes are therefore full of abstract thoughts and even seem a little sad. Yet Einstein is a very cheerful person among company. He loves jokes, smiles good-naturedly like a child. Then, for a moment, his eyes turn very childlike. His unusual simplicity creates an affinity, and so everyone will want to treat him amicably, shake his hand, pat him on the back—and all this naturally with the greatest respect. One somehow gets a feeling of tender sympathy, of recognition by a great defenseless simplicity, and at the same time a feeling of boundless respect. (22, p. 481)

The official meetings between Einstein and Lunatsharski also had a private component, because Lunatsharski's literary secretary, Dmitri Marianoff, and Einstein's stepdaughter Margot fell in love and got married in 1930. Marianoff was also employed by the Soviet Chamber of Commerce in Berlin. The marriage unfortunately did not last, as they separated after emigrating to France.

For all his sympathy with the Soviet Union, Einstein's relations with it were not always positive. But mutual gestures of respect between the scientist and the Soviets continued through the 1920s. In the spring of 1924, for instance, Einstein attended the founding of the German-Soviet community organization Kultur und Technik, which had its headquarters in Moscow, and in 1923 the Academy of Sciences of the USSR elected him as a corresponding member, subsequently raising him to honorary member in 1927. Later, after the first signs of Stalinist terror and when Einstein's theory of relativity had been denounced as bourgeois, relations cooled visibly. By 1929 Einstein had distanced himself from the Friends of the New Russia and protested Germany's refusal to grant political asylum to Stalin's rival Leon Trotsky when he was forced to emigrate. In 1932 he refused to add his signature to an appeal drafted by the French writer and Communist Henri Barbusse because it glorified the situation in a biased way. He wrote to Barbusse in this regard:

> I put much effort recently into making an informed opinion on the developments there and have arrived at quite grim results. At the top, personal battles by power-hungry persons using the most despicable means out of purely egoistic motives. Toward the bottom, complete suppression of the individual and freedom of expression. (22, p. 395; 47, trans. p. 257)

Einstein's critical attitude toward the political conditions in the Soviet Union softened under the impression of the high death toll for the Red Army and the immense suffering that the Soviet peoples endured during World War II. But when the personal cult around Stalin reached a new climax in the postwar period with its ideological controversies, Einstein's view of the world and politics, his pacifism, Zionism, and internationalism came under the scrutiny of Communist Party philosophers and ideologues. In a campaign against "cosmopolitanism" toward the end of the 1940s, Einstein was personally targeted, and his relativity theory was denounced as "idealistic," "reactionary," and "bourgeois." Leading Soviet physicists continued to acknowledge the universal validity of Einstein's relativity theory, though, and under the impression of its success in the building of the atomic bomb, the ideological agitators were never able to gain the upper hand in Soviet science and politics. Nevertheless, in 1952 Einstein dedicated the following sarcastic verse to these controversies over his theory (12, no. 31-418):

> Sweat and work away as never
> just a grain of truth to gather?
> Fool is he who must so labor;
> it's simply passed by party favor.
>
> And those who dare to doubt in it
> get their skulls cracked up a bit.
> Yes, that's a novel way to teach
> the bold in spirit harmony.

Durch Schweiss und Mühe ohnegleichen
Ein Körnchen Wahrheit zu erreichen?
Ein Narr, wer sich so kläglich schinden muss
Wir schaffen's einfach durch Parteibeschluss.

Und denen, die zu zweifeln wagen
Wird flugs der Schädel eingeschlagen.
Ja, so erzieht man, wie noch nie,
Der kühnen Geister Harmonie.

Site 18. New Synagogue
Now Centrum Judaicum
Oranienburger Strasse nos. 28–30
10117 Berlin (center)

met. train sta. Oranienburgerstrasse (S1, S2)
or Hackescher Markt (S5, S7),
from there 2 or 5 min., respectively, by foot

"I first discovered that I was a Jew when I came to Germany fifteen years ago, and this discovery was conveyed to me more by non-Jews than by Jews" (5, p. 104), Einstein wrote in 1929 to the journalist and former minister of education for the province of Baden, Willy Hellpach. Judaism, he said, had been buried for him since his youth. After a brief revolt during puberty, when at the age of twelve he rebelled against his parents' assimilated lifestyle and for a short time even demonstratively adhered to the religious practices, he became increasingly alienated from Judaism and developed into a "free thinker," as he once described himself with regard to religion. Even while in Prague, where he was employed as professor at the German University from 1911 to 1912, there was no profound reconsideration of his own Jewishness, despite living in the vibrant Jewish community there. Although he was personally acquainted with Hugo Bergmann and mingled with the active group of Zionists that philosopher led, Einstein was not particularly impressed with their discussions. In 1916 he wrote deprecatingly to Hedi Born about the "small circle there of philosophical and Zionist enthusiasts, which was loosely grouped around the university philosopher, a medieval-like band of unworldly people" (16, trans. p. 4).

Fig. 3.12. New Synagogue, 1920s

After settling down in Berlin and experiencing the outbreak of World War I, Einstein became politically aware; his sharpened political eye sensitized him increasingly to the social situation of Jews, whom he referred to as his "fellow clansmen" or "brethren" (*Stammesbrüder*). The virulent anti-Semitism prevalent in German society, indeed, even among academics, was not the only catalyst for this rediscovery of his own Jewishness. He had naturally encountered the same kind of prejudice in Prague. His "fellow clansmen" also gave Einstein plenty of opportunity to think about his Jewish identity. The strained assimilation attempts he was able to observe, even very close to home among his family and friends, he thought were shameful and despicable. It was particularly prevalent among the wealthy and educated Jewish bourgeoisie, or *Bildungsburgertum*. Einstein's diagnosis of this weakness was a lack of pride, self-esteem, and solidarity, particularly in their treatment of the growing problem of eastern Jewish migration, which particularly affected Jewish life in Berlin during and after World War I. In the years following that war, more than thirty thousand eastern European Jews were living in Berlin, about a fourth of the Jewish residents of the city. But only a minority of Berlin's successful Jews wanted to accept these brothers and sisters in the faith. Most were impoverished and still strongly attached to traditional lifestyles and rituals and therefore generally perceived as unpleasantly burdensome, to be avoided as far as possible; sometimes this attitude was even based on anti-Semitic stereotypes. The situation was made worse by an aggressive propaganda campaign by nationalists and racists aimed at expelling or confining eastern European immigrants. The Jewish Workers' Bureau asked Einstein in the fall of 1919 to state his position on this situation. Feeling this issue touched his sense of justice and solidarity, Einstein responded with an article in the *Berliner Tageblatt* of December 30, 1919. Under the heading "Immigration from the East" he objected to the slanderous agitation and made clear that "the recuperation of Germany truly cannot be accomplished by the use of force against a small and defenseless fraction of the population" (1, vol. 7, trans. p. 111); such inhuman measures were rather liable to squander Germany's moral capital and reputation.

This newspaper article was Einstein's first public statement on the Jewish issue, and his Zionist involvement followed in its wake. His first contact with that movement had been made in February 1919. Initially skeptical about the idea of Zionism, Einstein's attitude gradually changed, and his interest deepened in the years that followed (72b). Although he became an outright supporter, he always retained a high degree of independence and never officially joined any Zionist organization. He was constantly aware of the contradiction between the Zionist ideal and his internationalism. To this extent his loyalty was based only on cultural Zionism, which allowed him to live a Jewish identity safely away from any religious duty while remaining consistent in opposing assimilation trends. Einstein's conception of Jewishness had primarily mental, moral, and cultural components, and Zionism was "the only effort, which leads us closer to this goal" (30, p. 72) of reviving a sense of community.

Einstein's interest in Zionism was not confined to words of sympathy. Beginning in the 1920s he also offered his services to numerous propaganda and funding campaigns in support of the Zionist movement. Acting as a "bigwig and

Fig. 3.13. Albert Einstein and his secretary Helen Dukas in the New Synagogue on the occasion of his benefit concert in 1930

decoy-bird" (17, p. 40)—as he put it in a letter to his friend Maurice Solovine—he accompanied the Zionist leader Chaim Weizmann on a fund-raising tour of America in the spring of 1921 (61b). It was a great success for the Zionist movement as well as for Einstein himself. After returning home, he wrote Paul Ehrenfest in Leyden in June 1921 that he was "very happy to have responded to Weizmann's invitation" (30, p. 75).

In the *Jüdische Rundschau* of July 1, 1921, he even confessed that it was in America that he first discovered the Jewish nation:

> I've seen a great many Jews, but I never saw a Jewish people in Berlin or elsewhere in Germany. This Jewish people that I saw in America had immigrated to America from Russia, Poland, and Eastern Europe in general. These people still carry a healthy national feeling within them that has not yet been destroyed by the atomization and splintering of the individual. (1, vol. 7, trans. p. 244)

Jewishness never was a question of religious faith for Einstein but a matter of sharing in the same social destiny. He saw it as defined by a common history and culture in which were rooted certain ethical values and attitudes toward life. In the 1920s he professed, "A striving for knowledge for its own sake, a love of justice bordering on fanaticism, and a striving for personal independence—these are motifs of the tradition of the Jewish nation allowing me to feel my membership as a gift of fate" (5, p. 89).

Thus Einstein's Jewishness was formed during his Berlin years on the foundations of these convictions and suspended in the tension between his socialistic, cosmopolitan, and pacifistic ideals. It became the defining element of his worldview and influenced his political views and activities for the rest of his life.

Although Einstein felt attached by a common fate to his "fellow clansmen," quite some time was needed and much effort of persuasion on the part of the Jewish community before Berlin's most prominent Jew would acquiesce to becoming a member of their *Kultusgemeinde* (congregational association). In 1920 Einstein still rebuffed a demand by the community that he pay his delinquent community taxes with the gruff statement: "After careful consideration, I cannot resolve to join the Jewish religious community. As much as I feel myself to be a Jew, I stand aloof from the traditional religious rites" (42, p. 7; 1, vol. 10, trans. p. 338).

But the Jewish community of Berlin would not content itself with this verdict and persevered in the years that followed in its attempt to change Einstein's mind and enroll him as a member. They eventually succeeded. On February 24, 1924, he finally declared his willingness to became a taxable member of the community—but for him it remained purely a matter of conviction and solidarity and not any kind of religiosity. His consistent rejection of all religious regulations and institutions as well as his denial of the existence of a personal god barred him from this. He preferred to join his revered Spinoza in the belief that the only manifestation of divine reason is in the harmony of the natural laws.

Being a part of the Jewish community was of great value to Einstein just as much as it was binding. He met this obligation by lending his name and influence to the Zionist cause and thus became the most important and world-renowned symbol of this movement. Many other Jewish organizations were granted the use of his name and popularity for their purposes as well. He spoke at countless events organized by Jewish student organizations and accepted invitations to appear at highly publicized events to generate ticket sales on behalf of Jewish relief organizations. One of these was the benefit concert conducted by the Jewish Community of Berlin that took place on January 29, 1930, in the New Synagogue. Einstein appeared on that occasion as a violinist accompanied by the violinist Alfred Lewandowski, the vocalist Hermann Jadlowker, and the organist Arthur Zepke as well as the extended choir of the synagogue. Einstein played Handel's second sonata in B major and the adagio from the double concerto in C minor (transcribed for organ accompaniment) by his favorite composer, Johann Sebastian Bach. Einstein probably did not take part in the selection of compositions from the Jewish liturgy that followed. The profits from the well-attended

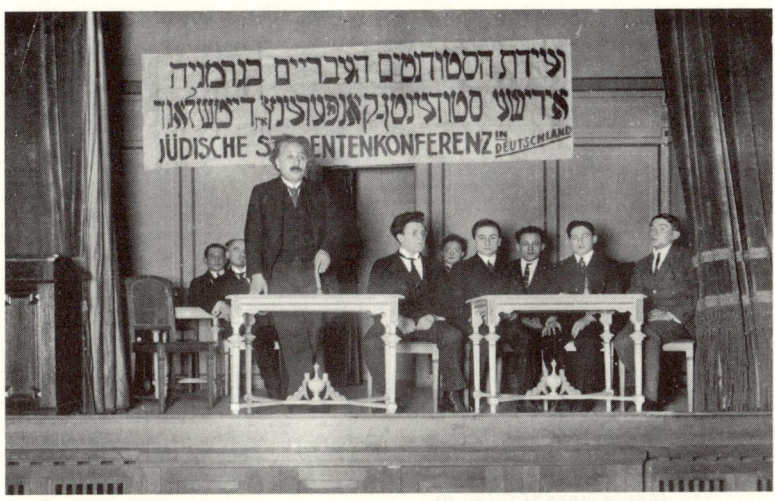

Fig. 3.14. Einstein speaking before the Jewish Student Conference, Berlin, 1924

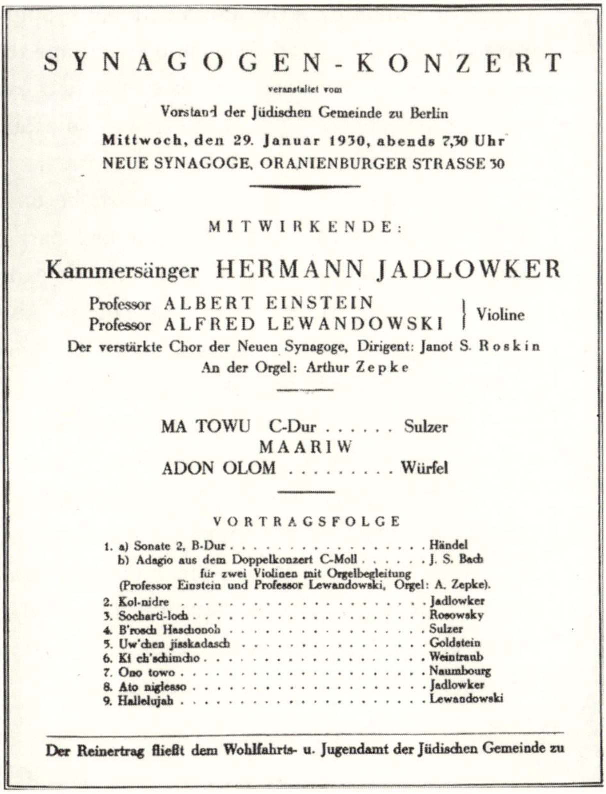

SYNAGOGEN-KONZERT

veranstaltet vom

Vorstand der Jüdischen Gemeinde zu Berlin

Mittwoch, den 29. Januar 1930, abends 7,30 Uhr

NEUE SYNAGOGE, ORANIENBURGER STRASSE 30

MITWIRKENDE:

Kammersänger HERMANN JADLOWKER

Professor ALBERT EINSTEIN } Violine
Professor ALFRED LEWANDOWSKI } Violine

Der verstärkte Chor der Neuen Synagoge, Dirigent: Janot S. Roskin

An der Orgel: Arthur Zepke

MA TOWU C-Dur . . . . . . Sulzer
MAARIW
ADON OLOM . . . . Würfel

VORTRAGSFOLGE

1. a) Sonate 2, B-Dur . . . . Händel
   b) Adagio aus dem Doppelkonzert C-Moll . . . . . . J. S. Bach
   für zwei Violinen mit Orgelbegleitung
   (Professor Einstein und Professor Lewandowski, Orgel: A. Zepke).
2. Kol-nidre . . . . Jadlowker
3. Socharti-loch . . . . Rosowsky
4. B'rosch Haschonoh . . . . Sulzer
5. Uw'chen jisskadasch . . . . Goldstein
6. Ki ch'schimcho . . . . Weintraub
7. Ono towo . . . . Naumbourg
8. Ato niglesso . . . . Jadlowker
9. Hallelujah . . . . Lewandowski

Der Reinertrag fließt dem Wohlfahrts- u. Jugendamt der Jüdischen Gemeinde zu

Fig. 3.15. Program for a concert featuring Albert Einstein playing violin, 1930

concert went to the benefit of the Welfare and Youth Bureau of the Jewish Community of Berlin.

Site 19. Exhibition grounds
Hammarskjöldplatz
14055 Berlin (Charlottenburg)

subway stops Kaiserdamm or Theodor-
   Heuss-Platz (U2) or
met. train sta. Messe Nord/ICC (Ring),
from there 10 or 5 min., respectively, by foot

In October 1923 German radio broadcasters aired their first regular programs, and in December 1924 the Radio Fair was brought to life in Berlin. Its purpose was to boost sales of radio sets and encourage the spread of this new medium of communication. Broadcasts of spectacular events such as boxing matches and other sports events achieved this to an extent. Under the banner "culture through technology" numerous German intellectuals, artists, and politicians of the time demanded that radio broadcasts become accessible to a broader audience. It is not known whether Einstein took part in this initiative, but he repeatedly made use of the new medium relatively early on. In 1929, for example, he congratulated the American inventor Thomas Alva Edison in a speech broadcast on the fiftieth anniversary of his invention of electrical lighting.

The organizers of the seventh German Radio Fair in 1930 invited Einstein to give the inaugural address at this exhibition of German radio technology on the exhibition grounds next to the Berlin radio tower. The inauguration took place on August 22, 1930, and with Einstein as the featured speaker, a popular and effective draft horse had been harnessed to attract crowds and turn it into a special occasion. His moody, now proverbial address "Esteemed present and unpresent audience!" (Verehrte An- und Abwesende!) testifies to the charged atmosphere along with the images and sound recordings taken during the event. The result was one of the early documentary films of Einstein.

> Esteemed present and unpresent audience! When you listen to radio broadcasts, also think about how humans came to possess this wonderful tool of communication. The original source of all technical achievements is divine curiosity and the

Fig. 3.16. Exhibition grounds at the foot of the radio tower, 1931

impulsive playful tinkerings of the pensive scientist, no less than the constructive imagination of the engineer inventor.

Think of Oersted, who first noticed the magnetic effect of electric currents; Reis, who first used this effect to produce sound by electromagnetic means; Bell, who by using sensitive contacts with his microphone first transformed sound frequencies into variable electric currents. Think also of Maxwell, who demonstrated by mathematical means the existence of electric waves; Hertz, who first produced and verified them with the aid of a spark. Consider especially also Lieben, whose idea to use an electrical valve tube revealed it as an unrivalled detector for electric frequencies that, at the same time, proved to be an ideally simple instrument for generating electric frequencies. Gratefully consider the army of nameless engineers who simplified the instruments of radio communications and adapted them to mass production to the point that they have become accessible to everyone. Everyone should be ashamed to use wonders of science and technology thoughtlessly without more understanding than a cow has about the botany of the plants she sumptuously devours.

Think also about how technicians are the ones to first make true democracy possible. For they not only ease people's daily chores but also make available to all the works of the finest thinkers and artists, which just a short while ago was a

Fig. 3.17. Einstein delivering his inaugural address at the Radio Fair of 1930

privilege of favored classes to enjoy, and thus peoples are roused out of their sleepy dullness.

As concerns radio broadcasting, in particular, it has a unique function to fulfill in the sense of reconciliation among nations. Up to our times, the nations learned about one another almost exclusively through the distorted mirror of their own daily papers. Radio presents them to one another in live form, and generally from their likable aspect. It will thus contribute toward rooting out the mutual feeling of strangeness, which so easily turns into mistrust and animosity.

In this state of mind regard the results of the creations which this exhibition offers to the astonished senses of its visitors. (11, supplementary issue, pp. 3–5)

The hope Einstein expressed in his speech that radio—and technology generally—would contribute to furthering democracy and have a reconciliatory effect on nations and consequently "root out the mutual feeling of strangeness, which so easily turns into mistrust and animosity," would very soon prove to be illusory—if, indeed, it wasn't already so at the very moment Einstein was speaking. Three years later, Joseph Goebbels opened the Radio Fair and made very clear the importance that radio was to have in National Socialist propaganda. This use of radio broadcasting had absolutely nothing in common with the hopes Einstein had aired about the medium, and Einstein himself became its favorite target.

CHAPTER FOUR

A Circle of Friends and Acquaintances

Describing Einstein as a "social lion" is surely misplaced and would contradict his own self-characterization as a "typical go-it-aloner in daily life" and "individualist" (*Eigenbrödler*). Even so, Einstein was firmly embedded in the social life of Berlin. Besides maintaining strong ties with his family, with closer scientist friends, and with girlfriends, he also kept contact with political, economic, and cultural figures. Invitations to a soirée or afternoon tea at Haberlandstrasse were quite regular; more frequently, though, Einstein went out to visit friends and colleagues, probably enjoying exquisite dinner invitations among high society. Einstein's politician friends and acquaintance included Walther Rathenau, Count Harry Kessler, and Gustav Stresemann. There was mutual appreciation not just on the personal level but also in others' awareness of Einstein's role as Germany's billboard and "cultural factor of the first order" in German foreign policy. His revived interest in Judaism brought him contacts with representatives of the Berlin Jewish community and particularly with representatives of the Zionist movement like Kurt Blumenfeld. His relations with the Berlin bankers Hugo Simon and Leopold Koppel were certainly owing to Einstein's role as the most important scientist of his day, especially considering that the latter was one of the patrons responsible for funding Einstein's appointment to Berlin. A relaxed if more distant friendship was entertained with the literary prince Gerhart Hauptmann and his family. There were other acquaintances among Berlin's intellectuals and artists as well—such as with the publisher Samuel Fischer, the chess champion Emanuel Lasker, and the architect of the Einstein Tower, Erich Mendelsohn. Many painters considered it a great honor (and no doubt good publicity) to be able to paint Einstein's portrait—the list ranges from Emil Orlik and Max Oppenheim to Leonid Pasternak and Max Liebermann.

There are a number of portraits by the latter, about which Einstein quite disrespectfully commented: "I thought that the picture resembled him more than me, which came in handy for him" (97, AE to C. Seelig, April 20, 1952). Einstein's passion for music and home performances drew him in closer contact with prominent musicians like the conductor Erich Kleiber, the violin virtuoso Fritz Kreissler, and the pianist Joseph Schwarz.

The focus of Einstein's social network in Berlin lay unquestionably with his scientist friends and colleagues: among these were the pioneer of wireless telegraphy known as the "Red Count," Georg von Arco; the editor of the *Naturwissenschaften*, Arnold Berliner; the president of the PTR, Emil Warburg; as well as his closer friends Fritz Haber, Max von Laue, and Max Planck.

All of these friendships are very well documented, so the following sketch provides only a few samplings. An equally striking characteristic of Einstein's friendships is the disproportionate number of medical doctors among them—including the surgeon Moritz Katzenstein, whom Einstein described as one of his closest friends, and the specialist in internal medicine János Plesch, as well as the physiologist and pacifist Georg Friedrich Nicolai, the radiologist Gustav Bucky, the internalist Rudolf Ehrmann, the neurologist Otto Juliusburger, the dentist Josef Grünberg, and the practitioner Hans Mühsam. Most of these closer personal relationships were maintained in American exile. Nevertheless, Einstein was anything but a hypochondriac, rather eying those of the medical profession and their art with fundamental skepticism: "I generally think little of this guild and find it better, on the whole, to spare Nature that vexation," he wrote to his friend Michele Besso in 1940 (18, p. 352).

Rudolf Ehrmann attempted to explain his attraction to doctors, at least as personal friends, with the following sociopsychological motives: "He probably liked conversations with medical doctors because doctors gain their personal experiences in the most disparate spheres of human society and not just within a small group as in most professions" (28, p. 229).

This explanation would at least agree with Einstein's general rejection of overspecialization, which had a role in his scientific breakthrough in his "wondrous year" 1905. In any case, Einstein "loved brainy characters" (22, p. 169), and these undoubtedly included the majority of his friends in Berlin besides the doctors already mentioned. He melancholically longed for this close-knit group of friends in a letter written in American exile in 1934:

> The small group of people, who were earlier so harmoniously attached to each other, really was unique and has hardly met its match again at this personal level

since. On the other hand, I do openly admit not to have shed a single tear about the broader circle of this acquaintance; for the uninvolved observer it was more amusing than it was lovable. (96, AE to M. von Laue, 23 March 1934)

Friends around Einstein

Site 20. Max von Laue
Albertinenstrasse 17
14165 Berlin (Zehlendorf)

subway stop Zehlendorf (S1),
from there, 15 min. by foot

"Laue wants to come here. . . . The poor fish. Nervous subtlety. To strive for an aim which is in direct contradiction with his natural desire for a quiet life, free from complicated human relationships. In this connection please read Andersen's pretty little fairytale about the snails" (16, trans. p. 5). With these hardly complimentary words in a letter to Hedi Born, Einstein characterized Max von Laue's efforts in December 1918 to swap his professorship at the University of Frankfurt for Max Born's at Berlin. Not a good basis for a friendship, one would think. And yet, after Laue had settled in Berlin—the exchange was worked out in 1919—a close relationship between trusted friends developed between the two scholars that survived even the years of the Third Reich unharmed. In Berlin Laue gained the confidence of his colleague Einstein in settling professional affairs, becoming one of his closest friends.

The two scientists had first met very early on. Having been made aware of the genius at the Berne patent office by his teacher Max Planck, Laue used the occasion of a tour of the Alps in August 1907 to make a detour to Berne in order to meet Einstein:

> Following our agreement by letter I looked him up at the Office of Intellectual Property. In the general reception area an official told me I should go out again into the hallway, Einstein would meet me there. I did so, too, but the appearance of the young man who walked toward me was so unexpected that I could not believe he could be the father of relativity theory. (20, p. 130)

Fig. 4.1. Residence of the von Laue family with Laue's Steyr parked in front, 1938

Laue was very quickly put right and became one of the early advertisers of relativity theory—initially in a series of papers and then, in 1911, in the first book describing Einstein's theory, *Das Relativitätsprinzip*; a second volume on the general theory of relativity followed in 1921.

Einstein respected Laue's work as well. He considered the patterns of X-rays diffracted by a crystal that Laue and his colleagues in Munich Walter Friedrich and Paul Knipping had developed to demonstrate the wavelike nature of X-rays in 1912 "the most wonderful thing I have ever seen! Diffraction off the individual molecules, thus revealing their arrangement" (1, vol. 5, p. 483).

When Laue moved to Berlin in the winter of 1918–19—less out of ambition than to be as near as possible to his teacher Max Planck—the professional acknowledgment and admiration between Laue and Einstein quickly turned into close friendship. An important element of this was not just Laue's early advocacy of Einstein's relativity theory but his public appearances at the beginning of the 1920s against the first wave of defamations and attacks on relativity theory. In the summer of 1920 he stood up for Einstein and his theories when

Einstein became the target of nationalistic and anti-Semitic attacks at the Berlin Philharmonic. In a press release he drafted with his fellow Berlin physicists Walther Nernst and Heinrich Rubens, Laue declared:

> Whoever has had the pleasure of knowing Einstein more personally knows that no one can outdo him in acknowledging the work of others, in personal modesty and in his distaste for publicity. It seems to us an obligation of justice to express this conviction of ours without delay, since no opportunity was given us to do so yesterday evening. (50, p. 59; 47, trans. pp. 101–102)

When two years later Einstein felt so personally at risk that he canceled his participation at the jubilee convention of the Society of German Scientists and Medical Doctors in Leipzig, Laue willingly took his place on short notice to speak about relativity theory. He demonstrated considerable personal courage by doing so, because protesters had come in force to distribute flyers and conduct vociferous demonstrations against the theory. This and the fact that at about the same time Einstein was engaging Laue increasingly in the management of the Kaiser Wilhelm Institute gave more opportunities for the two scholars to work together and become closer friends. During the 1920s they met frequently not just on scientific and social occasions at the academy, the Kaiser Wilhelm Society, the university, and other scientific facilities of the city but also privately. Their two families got together at Haberlandstrasse in the Schöneberg district and at Albertinenstrasse in Zehlendorf, where Laue had bought himself a home at the beginning of the 1920s with the proceeds from his Nobel Prize award. The Laue family lived there until 1944, when Berlin's Kaiser Wilhelm Institutes and their staffs were evacuated to southern Germany because of the bombing raids on Berlin. What the maid of the Einstein household said about Laue being among the most frequent visitors to Haberlandstrasse probably also holds true for Einstein at Albertinenstrasse. The frequency of their mutual visits is not the only barometer by which to gauge the strength of their friendship. Konrad Wachsmann said that Laue was the only person Einstein felt relaxed enough to "joke around" with. The following anecdote is revealing:

> When Laue had done or settled something for Einstein, he asked at the end: "Isn't that fine of me?" Einstein replied: "I did tell you many times that the finest thing about you is your wife, then comes the diffraction pattern, and last of all your magnificent mane." . . . The mischievous humor between Einstein and Laue was so contagious that roaring laughter almost always accompanied Laue's visits to Haberlandstrasse or Caputh. (22, p. 173)

A photograph of one of Einstein's return visits to the Laues on Albertinenstrasse has survived. Einstein and Laue are sitting together with Max Planck and Walther Nernst and the visiting American physicist and Nobel laureate Robert Andrew Millikan. Such a high density of Nobel laureates around a living room table would be unusual even today, particularly in Berlin.

This picture focuses like a magnifying glass on the stature of the city at that time as a world center of scientific research, specifically in physics, and documents the loss that Berlin experienced as a result of the National Socialist's persecution of Jews after 1933. Even though Laue thought Einstein's statements about the Nazis' rise to power had been inappropriately political for a scientist, he remained Einstein's advocate during this period. His letter from May 1933 reveals how deeply Einstein's departure affected him: "Your leaving here has largely devastated Berlin for me; despite Planck, Schrödinger, and some others" (12, M. von Laue to AE, 14 May 1934).

Laue was one of the few German scholars to stand up for their convictions during the period of the Third Reich and—to use Einstein's terms—stay morally upright. Many physicists abroad agreed with Einstein, acknowledging Laue

Fig. 4.2. Gathering of Nobel laureates at Laue's residence, 1931. *From left to right:* Walther Nernst, Albert Einstein, Max Planck, Robert Millikan, Max von Laue

as a "knight without fear or fault." In the postwar years his moral rectitude and scientific reputation helped renew international ties with German physicists and promote the restoration of physics in Germany. Despite many attempts, not even Laue managed to motivate his old friend and colleague to visit Germany or particularly Berlin. Laue's final attempt to invite Einstein in the name of physicists of the East and West to attend the celebrations of the fiftieth anniversary of his annus mirabilis in divided Berlin, received the friendly but resolute reply:

> Above all, I am glad that in this unusual case I provide reason for brotherly collaboration and not for controversy.... Age and sickness make it impossible for me to attend such occasions and I must also admit that there is something liberating about this divine providence. For everything that is in anyway related to a personal cult has always been irritating to me. (59, p. 88)

Site 21. Max Planck
Wangenheimstrasse 21
14193 Berlin (Grunewald)

subway stop Halensee (Ring),
from there, 10 min. by foot

"Planck loves you," Elsa Einstein wrote in summer 1921 to her husband, pointing out the special relationship between Albert Einstein and Max Planck. It would be no exaggeration to call Planck Einstein's "discoverer," as he was among the first prominent physicists to recognize the revolutionary significance of Einstein's papers from 1905, particularly his theory of relativity, and to spread word about it. As early as the spring of 1906 Planck gave a report on Einstein's paper "On the Electrodynamics of Moving Bodies" at the physics colloquium in Berlin, and in the years that followed, he encouraged most of his doctoral students to examine issues related to relativity theory, not quanta, despite being the rather reluctant "father of quantum theory." At the Scientists' Convention of 1906 he even defended Einstein's theory when Walter Kaufmann presented experimental results that appeared to contradict the relativistic stipulation that the mass of an electron depends on its velocity. In 1913 Einstein gratefully acknowledged in his essay "Max Planck as Scientist": "It is in great measure due to the decisiveness and warmth with which he [Planck] championed this theory that fellow physicists began so quickly to pay attention to it" (1, vol. 4, trans. p. 274).

Fig. 4.3. Planck's villa in Grunewald, Berlin, 1910; destroyed in World War II

Planck was also among the most determined supporters of Einstein's appointment to Berlin, comparing Einstein's scientific achievements with those of Copernicus. Although the two scholars could not have been more different in their personal and political views, their appreciation and respect for each other deepened as time went by. Planck "is a splendid person," Einstein wrote in December 1915 to his friend Besso (1, vol. 8, trans. p. 163); Marga Planck reciprocated in her letter to Einstein from 1918, expressing her joy "that my husband has found such a warm friend in you!" (1, vol. 8, trans. p. 545). This mutual appreciation went beyond acknowledgment of scientific merit and professional competence. They shared a profound personal sympathy. Einstein was deeply moved by Planck's personal losses during World War I—his son Karl fell at Verdun, and both his twin daughters died in childbirth. For Einstein, too, it was a time darkened by deep personal crisis. On December 1, 1919, he wrote to Moritz Schlick about visiting Planck just after he had lost his second daughter: "Yesterday I visited Planck, without being able to hold back my tears at the sight of him. He was in command of himself and composed—a truly great and exceptional person" (1, vol. 9, trans. p. 157).

Planck was a passionate music lover, and played the piano with the perfection of a professional musician himself. So he frequently invited his friends and colleagues to his villa on Wangenheimstrasse in the Berlin district of Grunewald to play music. One guest at these social gatherings was Einstein, who came with his violin. Planck and Einstein were not the only music makers; occasionally Planck's son Erwin, who played the cello, joined them to form a trio. Lise Meitner wrote of one of these musical evenings at the Planck home in a letter to Otto Hahn in the fall of 1916: "Yesterday I was visiting the Plancks. Two magnificent trios (Schubert and Beethoven) were being played. Einstein was playing the violin, emitting all the while such pricelessly naive and peculiarly political and warlike quips" (96, L. Meitner to O. Hahn, November 16, 1916).

At the beginning of the 1920s Einstein received various attractive job offers abroad and was thus repeatedly confronted with the question of whether or not to turn his back on Germany—motivated by the ever more strident attacks on his theory and his person. It was apparently Planck who always managed to convince him to stay in Berlin. When renewed assassination threats in the summer of 1923 caused Einstein to flee to Leyden, Planck feared he would stay there permanently. In a letter he begged Einstein not to act precipitously: "I am quite beside myself with fury and anger about these infamous obscurantists, who have dared and managed to separate you from your home, from the site of your activities" (12, M. Planck to AE, November 10, 1923).

Einstein came back, probably not least because—as he revealed in a letter to his friend and colleague at Leyden, Paul Ehrenfest as early as 1919—he had promised Planck:

Fig. 4.4. Erwin Planck, Albert Einstein, and Max Planck at a reception in the Reich Chancellor's palace, 1931

I won't turn my back on Berlin unless conditions arise that would make such a step seem natural and proper. You can scarcely imagine what sacrifices are being made here in the difficult financial situation overall to make it possible for me to stay, and to maintain my family in Zurich! It would be doubly repulsive of me if, just at the moment when my political hopes are being realized, I turned my back, without good reason, perhaps in part for the sake of material advantage, on people who have shown me affection and friendship and to whom my leaving at this so-called dishonorable time would be doubly painful. . . . I can go away from here only if there is a turn of events that makes it impossible for me to continue to stay. Such a change could occur. If it doesn't happen, my departure would be equivalent to a shameless breach of faith toward Planck and disloyal in any case. I would have to reproach myself for it later. (1, vol. 9, trans. pp. 86–87)

The rise to power of the National Socialists at the beginning of 1933 brought such a turn of events. Einstein refused to live in a country in which civil liberties and rights were being so broadly and severely violated as in Nazi Germany. He no longer felt bound by his promise to Planck. He was dismayed by the opportunism of many of his colleagues and at least as disappointed by Planck's own ambivalence toward Einstein's political behavior. Einstein could not comprehend Planck's maneuvering before the National Socialist leaders and their blatant acts of brute force and flagrant breaches of the law. He could not accept Planck's willingness to retain his position as a political authority—justifying doing so as an effort to save German science—and thereby allow himself to be used by the National Socialists. "Even as a Gentile [*Goj*] I would not have remained president of the Kaiser Wilhelm Society under such conditions" (12, AE to L. Silberstein, September 20, 1934), Einstein averred to an American colleague in the fall of 1934. Shortly before, in a letter to Fritz Haber, he had tersely rated the nobility of Planck and Laue: "Planck 60% noble and Laue 100%" (12, AE to F. Haber, 8 Aug. 1933).

This 60 percent high-mindedness still distinguished Planck from many of his German colleagues. Although their contact ended, Einstein evidently was able to hold his fatherly friend and mentor in some esteem even after the war and despite the Holocaust. In his epitaph to Planck that was read at the memorial session of the American Academy of Sciences in Washington, D.C., he remarked "that even in these times, in which political passion and raw violence inflict such great worries and suffering on people, the ideal of judiciousness is still upheld undiminished. This ideal . . . was embodied to rare perfection in Max Planck" (10).

Einstein's letter of condolence to Planck's widow gives a hint at what his meetings with her husband and their years together in Berlin meant to him:

> It was a fine and productive time that I was allowed to witness in his vicinity. . . . The hours that I was granted in your home and the many conversations I conducted in private with that wonderful man will remain among my finest memories for the rest of my life. The fact that a tragic fate tore us apart cannot change it. (12, AE to Marga Planck, November 10, 1947)

Site 22. Emanuel Lasker
Aschaffenburger Strasse 6a
10779 Berlin (Schöneberg)

subway stop Bayerischer Platz (U4, U7),
from there 3 min. by foot

In contrast to many other scientists and intellectuals, Einstein did not indulge in a passion for games—music and sailing were his main pastimes during his free time. He disliked chess in particular because he found it a far too "aggressive game, and I don't like that kind of fighting. But the main reason why I don't like chess is ethical: because the main goal of the game consists in beating the opponent through employing various tricks and stratagems" (38, p. 213).

This dislike did not, however, prevent Einstein from making friends with the world chess champion of the time. Emanuel Lasker shared common philosophical and epistemological interests, but perhaps their engagement in the Zionist cause also played a role. Lasker was born on Christmas Day in 1869 and was the son of a Jewish cantor. The name of this trained mathematician has gone down in the history of his field with the "Lasker decomposition principle," a generalization of his mathematical investigations. Despite such credentials Lasker was unable to follow an academic career; he devoted his efforts instead to professional chess playing—without entirely abandoning his mathematical and philosophical interests, as he continued to publish in these fields. His chess playing earned him some prosperity and led to extraordinary achievements. In 1894 he won the world championship, which he retained until 1921. He is not just the only German world champion in chess but to this day is the longest bearer of the title.

Lasker took up residence in Berlin in 1908, and he lived virtually "around the corner" from the Einsteins, at Aschaffenburgerstrasse no. 6a, in the 1920s. They

Fig. 4.5. View of Haberlandstrasse and Aschaffenburgerstrasse, 1920s

purportedly first met at the home of Einstein's friend and first biographer Alexander Moszkowski—probably in the fall of 1918, as Einstein wrote to his mother around that time:

> Recently I made the acquaintance of the chess master Lasker, a small, fine little man with a sharply cut profile and a Polish-Jewish, yet genteel manner. He has been world champion in chess playing for 25 years and is a mathematician and philosopher to boot. He stayed contentedly seated until 12 o'clock, even though a great tournament awaited him the next day. (1, vol. 8, trans. p. 664)

They became more closely acquainted—according to Einstein—in walks together, "during which we exchanged our opinions on a great variety of issues. It was a somewhat one-sided exchange, in which I was more the taker than the giver; for it was mostly more natural for this eminently productive person to develop his own thoughts than to adjust to those of someone else" (84, p. 3).

It hardly bothered Einstein that Lasker was a critic of the theory of relativity. Even on this point he found "Lasker's unwavering independence" to be "such a rare quality in mankind, among whom almost all, even intelligent people, belong to the class of conformers" (86, p. 4).

Einstein offered the following words (which might also be said of himself) in honor of Lasker's sixtieth birthday in December 1929:

Emanuel Lasker is one of the strongest minds I ever met in my life. A Renaissance man, gifted with an untamable urge for liberty; averse to any social bonds.... As a genuine individualist and self-willed soul, he loves deduction; and inductive research leaves him cold.... I love his writings, irrespective of their content of truth, as the fruits of a great original and free mind. (12, no. 28-060)

Besides a common love for long walks in deep intellectual conversation, Einstein and Lasker also cherished the environs of Berlin. Like Einstein, Lasker also moved out of the city in the summertime. He likewise had built himself an avant-garde house in Thyrow near Ludwigsfelde in Brandenburg. Whether the two scholars exchanged visits in their respective refuges in Thyrow and Caputh is not known.

Like Einstein, Lasker chose to leave National Socialist Germany in 1933; he first went to the El Dorado of chess, the Soviet Union, later to spend the last five years of his life in America. He died in New York in 1941.

There was apparently some contact between the two even in American exile. An exchange of letters survives in which Lasker asked Einstein to write a preface to his book *The Community of the Future*. Although the book centered on Lasker's political views, which had not changed significantly since the 1920s, Einstein refused his request. He acknowledged that the book "contains much wisdom ...

Fig. 4.6. Emanuel Lasker, around 1900

[yet] my views diverge on such important points from the ones you argue that I cannot support the book with a good conscience" (12, no. 53-735).

Another reason for this refusal must also have been that Einstein had meanwhile become a public personality. Beginning in the 1930s, particularly in the United States, with its ubiquitous mass media, he had to be very careful in his responses to such inquiries and requests. As he wrote just a short time later to his inventor friend Rudolf Goldschmidt in reply to a similar request, "Any activity outside my own field only draws ugly 'publicity' that absolutely must be avoided." It is not known whether Einstein and Lasker ever met again in exile to take up their Berlin discussions and perhaps even settle their differences. Be that as it may, ten years later, in 1952, Einstein did not turn down a similar request by Lasker's biographer, Jacques Hannak. He wrote a cordial preface to his biography in which he characterized the chess champion as "one of the most interesting persons I came to know in my later years" (55, p. 2).

Einstein did have contact with a distant cousin of Lasker's during Einstein's his American exile. Eduard Lasker, likewise a mathematician and chess player, had emigrated to the United States in 1914. Not only had he earned a name in international chess tournaments, but he was also an enthusiastic player of the Japanese board game go. Einstein probably liked this game even less than chess because the similarity to military tactics is even greater. So it might not have been chance that the go textbook Lasker had given to Einstein during his visit to Princeton appeared in a secondhand bookstore in Baltimore a short while later. Lasker was not terribly devastated to learn this, as he admitted himself that he had left Einstein's book on relativity theory in the New York subway.

Site 23. Moritz Katzenstein
Ahornallee 10
14050 Berlin (Charlottenburg)

subway stop Theodor-Heuss-Platz (U2),
from there 5 min. by foot

"Throughout the eighteen years I have been living in Berlin, few men have become close friends of mine, Professor Katzenstein was the closest" (20, p. 46), Einstein wrote in an epitaph to his friend in January 1932. Who was this Moritz Katzenstein, this regular guest at the Einstein home who probably welcomed

Fig. 4.7. Katzenstein's villa in the Westend district of Berlin, 1920s

Einstein as frequently in his own apartment on Ahornallee in the Westend district? By his own account, Einstein spent "the vacations of the summer months, mostly on his graceful sailboat" with Katzenstein over a period of "more than ten years," adding, "What we were experiencing, striving after, feeling, was shared there" (20, p. 46).

Born in 1872 in Rotenburg near Fulda, Katzenstein was one of the most renowned surgeons in Berlin during the 1920s—he directed the surgery clinic at the municipal hospital Friedrichshain from the beginning of that decade. After passing his final examinations in Wiesbaden, he studied medicine in Freiburg and Munich, where he took his doctorate in 1895. Following his military service he went to Berlin to train under the eminent surgeon James Israel at the Jewish hospital on Auguststrasse. There he also took an interest in the scientific analysis of medical problems, which remained a lifetime preoccupation of his, as his almost one hundred publications in the field attest. They treat arterial collateral circulation, testing of heart function, and the development of stomach ulcers. Some operational methods in trauma surgery and joint surgery are also attributable to Katzenstein. He was the first surgeon in Germany to stitch a six-year-old girl's torn meniscus back in place in February 1900. Despite skepticism on the part of many of his fellow surgeons, this success made him a pioneer of the treatment. His habilitation thesis on the arterial collateral circulation of the kidney at the University of Berlin in 1911, where he was appointed untenured professor of surgery in 1921, qualified him for academic teaching. He worked in various military hospitals at the front during World War I, after which he became—as his staff file puts it—"directing doctor at the hospital in Friedrichshain" (i.e., head doctor of the clinic). Bermann Fischer, who was an assistant doctor under Katzenstein in 1923 and later became a publisher, reminisced that "he could not have wished for a better training."

Einstein had not, of course, trained under Katzenstein, but he valued his consultations and his medical research. In his epitaph to his friend—Katzenstein had not even reached sixty when he died on March 23, 1932—Einstein wrote, "this friendship was animated not just by one understanding the other, by being enriched by the other and finding the resonance so indispensable to every truly living soul; this friendship also contributed toward making us both more independent from external experience, making it more easy to objectivize" (20, p. 46).

Fig. 4.8. Albert Einstein and Moritz Katzenstein sailing, around 1925

Site 24. János Plesch
Villa Lemm
Rothenbücher Weg
14089 Berlin (Gatow)

subway stop Theodor-Heuss-Platz (U2) or
met. train sta. Heerstrasse (S9, S75),
then by bus (lines X34, X49, or 134)
outbound to the town limit of Gatow,
from there 7 min. by foot

"From time to time now I sit all by myself in an apartment on a country estate and cook for myself—like the ancient hermits. Then one notices with astonishment how nice and long the day is and how superfluous is a large part of the busy or idle strivings in which one is wrapped up for the rest of the time" (18, p. 240), wrote Einstein in January 1929 to his friend Michele Besso after he had returned to the luxurious country retreat of his friend János Plesch in Gatow by Berlin. In this refuge Einstein found the necessary peace for concentrated work, as in the winter of 1928–29 when he completed his first draft of a unified field theory. Until he had his own house in Caputh, Einstein also used Plesch's property as a summer residence and as an escape from his own family and other turmoil that occasionally descended upon him, such as on his fiftieth birthday. Staying in Gatow for a few days was a good way to sidestep the onslaught of congratulators and well-wishers. He reportedly was usually lodged in a garden pavilion directly by the water and, to the delight of passing boaters, occasionally also played music.

János Plesch, born in Budapest in 1878, was a noted specialist in internal medicine known for his methodological advances in clinical analysis. He started working in Berlin in 1903, first at the Second Medical Clinic of the Charité under Friedrich Kraus, where he earned his habilitation degree and was described by Theodor Brugsch as one of his "most amusing colleagues at the clinic." In 1921 the University of Berlin conferred him the title of professor. Besides working at the Charité, Plesch ran his own profitable private practice, which in the 1920s attracted Berlin's rich and famous. His successful medical practice was not the only way he gathered such personalities around himself as others would collect postage stamps. The illustrious evenings held in his luxurious city residence by the zoo, a palace at Budapester Strasse nos. 22/23, were known throughout Berlin. His prominent guests from science, politics, and the arts included Fritz Haber, the diplomat Count Rantzau, the painter Max Slevogt, the musicians Arthur Schnabel and Fritz Kreisler and—last but not least—Albert Einstein.

When Einstein fell seriously ill with a cardiac problem in 1928 and—as he wrote his friend Besso—was "close to kicking the bucket," he exchanged Plesch's living room for his consulting room. But that caused some irritation and even strong protest, because Plesch's insatiable need to prove himself and his sometimes questionable manner of diagnosis and therapy had "very often caused ill-feelings among reliable doctors," including Einstein's own doctor friends. Einstein wrote to Heinrich Zangger in May 1928: "Ehrmann absolutely rejects Plesch, mainly for personal reasons. . . . But nobody is a flawless angel, so easy-does-it vis-à-vis all the other piggies" (12, AE to H. Zangger, May 5, 1928).

Plesch's autobiography, *János Tells about Berlin*, which also appeared in German translation in 1949 as *Janos erzählt von Berlin*, offers the reader a portrait of this fascinating character but also suggests the cause of the antipathy toward

Fig. 4.9. Villa Lemm, 1970s

Fig. 4.10. Albert Einstein and János Plesch, around 1929

him among his colleagues. These memoirs are amusingly written, but many of the episodes cannot be historically verified. The friendship between Plesch and Einstein in any case lasted until Einstein's death in April 1955. Plesch, who had also been forced to emigrate from Germany, was one of the last visitors at Einstein's home on Mercer Street in Princeton.

Women around Einstein

"He wasn't indifferent to women. He downright courted them," writes Claire Goll (82, p. 100), affirming that there were more women in Albert Einstein's life than his mother and two wives, Mileva and Elsa. His romantic relationships as a youth and a student reflect Einstein's inclination to woo the "weaker sex" clearly enough, and even in his later years amorous escapades and affairs were common. As a public personality, particularly in Berlin, this world-famous and charismatic professor attracted women—according to Konrad Wachsmann—"like iron filings to a magnet." Dmitri Marianoff, Einstein's son-in-law, remembered: "Many women tried to come into his life. Some wrote letters recalling to him a brief transitory meeting; others brought flowers and left them with a note and their addresses. We, of the household, understood these to be only the worthless emphases of fame" (29, p. 205).

Einstein's acquaintances with women were not always as platonic or meaningless as this description would lead us to believe. Einstein's Berlin years include at least four romances—disregarding the one with his cousin Elsa, which led in the end to their marriage. Initially it was the Austrian Betty Neumann, who had been engaged as a secretary after Einstein's return from his extended trip to the Far East. She was a niece of the doctor Hans Mühsam, who had treated Einstein's mother as well as Einstein himself. We know about Einstein's passionate liaison with Betty from remarks in correspondence with others, such as a letter to Hans Mühsam in which he mentions being "in raptures about your niece" and a love letter to her from Einstein from January 1924 (the remainder of his correspondence with her is probably still in safekeeping at the Einstein Archive in Jerusalem). In this letter he wrote his "Dearest Betty":

> I certainly would be happy if I knew you were somehow nearby and would sometimes be allowed to see your dear smile. But fate is merciless even toward the much envied such as I am. Because I'm not allowed to run after you, I do always hope to meet you by chance like that, but that happens so rarely, and then it's not even by chance. Dear Betty, laugh about me, old donkey, and find yourself a man who is ten years younger than me and loves you as much as I do, but do accept a kiss from your A. Einstein. (45, p. 135)

In the summer of 1924 the relationship between the two cooled noticeably. Whether Elsa's frequent jealous outbursts and monitoring marred their relationship or there were other reasons their relationship deteriorated is unknown. In any case, Betty's employment contract was not extended, and he wrote her that he now "had to seek in the stars what was denied to him on earth" (45, p. 206).

From Vienna in fall 1924 he even asked his friend and confidant Moritz Katzenstein to mediate in this touchy affair, expressing his doubts about

> whether I've done the right thing to treat little B.N. the way I have. Human values are of such a peculiarly subtle nature and cannot be measured by the pound. The

question is what someone has to lose if in reality he marries later on or does not do so at all and instead only has a loose relationship that is still very important to him. What I personally have come to know about marriage in my own case and others does not make me rate it very highly. Cross my heart! When one steps outside a bit, one does perk up. (AE to M. Katzenstein, Vienna, September 22, 1924)

In the years that followed Einstein held to his own maxim, and a graceful and attractive widow soon frequently appeared by his side. Her name was Toni Mendel, whom Einstein must have become acquainted with, along with her husband, through the New Fatherland League during World War I. From 1925 on, an intense friendship seems to have developed that involved the entire family. But Elsa probably only reluctantly respected it—despite the fine chocolates and other gifts Toni had intended for her. Until the year she emigrated in 1932, she frequently accompanied Einstein to concerts and operas. She usually came to Haberlandstrasse with a chauffeur to pick him up, despite Elsa's repeated scenes. The housemaid Hertha Waldow described a row Einstein and his wife once had about how little pocket money he had, paraphrasing him as saying that "if he was already accepting theater tickets as a gift and was being picked up by car, he really would like to have at least enough money on him to be able to pay Toni's and his cloakroom fees" (51, p. 44).

Fig. 4.12. Toni Mendel, undated

Fig. 4.11. Am Sandwerder, 1920s

Site 25. Toni Mendel's villa
Am Sandwerder 31
14109 Berlin (Wannsee)

subway stop Wannsee (S1, S7),
from there 10 min. by foot

One of the regular guests at Toni's lakeside "millionaire's villa" at Friedrich Karl Strasse no. 18 (now Am Sandwerder no. 31) was Einstein. He supposedly was up as early as six in the morning, waking up the other inhabitants of the house with his piano playing in the living room with its "wonderful view" on the Wannsee. The Mendel property, which comprised a number of parcels of land along the lakeshore, also included a private research laboratory of Toni's son-in-law, Bruno Mendel. Einstein took an active interest in his biochemical analyses of cancer. Einstein's biographer Friedrich Herneck reports that during his visits at Wannsee, Einstein would often perch himself "on a laboratory stool, admire the precision instruments and also give technical suggestions, for example about the design of an electrically heated bath needed for killing off cancerous cells by 'heat therapy'" (51, p. 147).

Toni Mendel was apparently not only physically attractive to Einstein but also an intelligent partner. According to Friedrich Herneck, they held intense conversations on philosophical issues. She was also the intermediary between Einstein and the Indian poet Rabindranath Tagore—at least one of their two meetings took place in the Mendel villa. Both also had a predilection for doggerel, as the following rhyme by Toni demonstrates:

A true philosopher, I am,
that's why I court Einstein.
Buy him everything there is,
that an old sinner loves.

Bin ein wahrer Philosoph,
Drum mach' Einstein ich den Hof.
Kauf ihm alles, was es gibt,
Was ein alter Sünder liebt. (27, p. 327)

Their acquaintance can be traced back to the early days of Einstein's residence in Berlin, and their friendship remained intact during his period as an émigré. Toni Mendel spent her final years in Hamilton, Canada, and they corresponded regularly even then. According to the Nobel laureate in chemistry Walter Kohn, their letters were "full of personal warmth and originality as well as sparkling wit" (58). Unfortunately this correspondence was destroyed after Toni Mendel's death.

There was another lover in Einstein's life at the beginning of the 1930s: Margarete Lebach, likewise an Austrian, who knew how to beguile not only Einstein but also Elsa with home-baked Viennese delicacies and even to shoo her away when she came on her weekly visits. As Herneck noted, "When she came, Frau Professor would always go into Berlin to do some errands or other business. She always went off into the city early in the morning and only came back late in the evening. She left the field clear, so to speak" (51, p. 44; trans. 29, p. 208).

If Einstein went to concerts and the opera with Toni Mendel, he invited Margarete Lebach out sailing in Caputh:

Dear Ms. Lebach!

I'm ashamed to have my magnificent sailboat if no one besides me gets something out of it. But most of all I'd love to be able to sail you around. So: No more miserable excuses! (22, p. 305)

Fig. 4.13. Albert Einstein sailing with an unknown woman, undated

The women Einstein was evidently attracted to were not just beautiful but well to do, as his fourth lover in Berlin also had a wealthy background. Estella Katzenellenbogen was the proprietress of a number of floral shops, whose chauffeur likewise picked Einstein up in her fine limousine for concerts and theater performances. Estella was the ex-wife of the general director of the Ostwerke, a mixed combine, whose income allowed her to maintain a large house near the Berlin zoo (Bendlerstrasse no. 40) after the divorce. Here, too, virtually all traces of the relationship have been erased, and we know very little about it—not even whether the following letter was addressed to Estella or another lover:

> Dearest. . . . I'm writing this letter at great risk, because Elsa could come in any moment, and that's why I really have to be careful. . . . Yesterday was so wonderful

that I'm still full of raptures. . . . I'm coming again to the same place at 5 o'clock, or better still, 10 minutes before 5, if you can arrange it. . . . Kisses to you, my love, from your A. E. (38, p. 235)

The jealous scenes are not the only evidence of how injured and humiliated Elsa felt about Einstein's affairs. Shortly before her death she wrote an American acquaintance of the family: "I think that you are the most considerate, loving husband. How gladly would I send Albert to you to be taught" (29, p. 215).

The Berlin Family

Berlin's unique research landscape and atmosphere were not the only factors determining Einstein's acceptance of the offer made by its leading physicists to move to the Prussian capital in the spring of 1914. The allure of his cousin Elsa Einstein, with whom he had recently fallen in love, also drew him to the city. It was certainly not without family connections for Einstein. His sister, Maja, had moved to the banks of the Spree as a young student to complete her first four semesters in the Romance languages—as she writes in her résumé. At that time university study presented major hurdles for women, and they had to be content with merely auditing courses—women were legally admitted to academic study at Prussian universities only in 1908. Maja registered at the University of Berlin as early as 1906 all the same. There she found in Adolf Tobler an internationally acknowledged teacher in the Romance languages who had likewise come to Berlin from Switzerland. Maja returned to Switzerland in 1907 to complete her doctorate at the University of Berne in 1909.

We do not know where Maja was living when she first moved to Berlin. But it is possible that relatives offered her bed and board. Her maternal cousin, Elsa Einstein, Albert's future second wife, had been married to Max Löwenthal, a businessman in the textile industry, in 1896 and was living in Berlin. Elsa's parents were also there. Rudolf Einstein, who was a cousin of Einstein's father, and his wife, Fanny, who was his mother's sister, had operated a textile factory in Hechingen in Württemberg before settling in Berlin in 1909. Rudolf was first listed in the Berlin directory as a businessman and later as a pensioner at the address Haberlandstrasse no. 5—the building in which their daughter Elsa also lived with her two daughters after her divorce from Max Löwenthal and likewise became the address of Albert Einstein in 1917.

Site 26. Jakob Koch
Wilmersdorfer Str. 93
10629 Berlin

subway stops Wilmersdorfer Strasse or
    Adenauer Platz (U7)
or met. train sta. Charlottenburg (S5, S7, S9),
from there 5 min. by foot

There were other maternal relatives living in Berlin besides Rudolf and Fanny Einstein. Jakob Koch, the brother of Einstein's mother, Pauline, had moved to Berlin in 1901 from southern Germany to seek his fortune in the grain trade. He lived in the elite suburb of Charlottenburg at, among other addresses, Mommsenstrasse no. 10 and, from 1908, Wilmersdorfer Strasse no. 93 (the house was destroyed in World War II). When Einstein first arrived in Berlin in 1912, he took up lodgings there. He lived there later repeatedly, especially when his mother Pauline kept house there for her brother for a year after the death of her sister-in-law in 1914. In 1916 she returned to her native southern Germany only to come back to Berlin fatally ill at the end of 1919 to spend her last days with her son in his apartment. She was buried in the cemetery in Schöneberg on Marxstrasse (now Eisackstrasse), but her grave no longer exists, as that part of the cemetery was demolished during construction of the Berlin throughway in the 1970s.

Fig. 4.14. Wilmersdorfer Strasse, corner of Giesbrechtstrasse, around 1900

Fig. 4.15. Pauline Einstein with her son, undated

If the strong presence of his kith and kin in Berlin served as an additional attraction for Einstein to brave the "Berlin adventure" despite all his reservations, it was a deterrent for Mileva. Her notoriously bad relations with her mother-in-law were compounded by her husband's relationship with Elsa, of which Mileva was well aware, and by the considerable pressures on their marriage imposed by the presence of an extended Jewish family in general. As Einstein would write to Elsa in December 1913, "My wife whines incessantly to me about Berlin and her fear of the relatives. She feels persecuted and is afraid" (1, vol. 5, trans. p. 371).

Her fears turned out to be justified. The move to Berlin spelled the end of her marriage, and she endured living there for only a few weeks; in July 1914 she returned Zurich, taking the children with her.

Einstein himself did not turn his back on Berlin until twenty years later, when he and his family were forced into exile by the Nazis. Einstein's emigration ended not only the Berlin chapter of his life but also the city's relations with the entire Einstein family. Elsa's parents and her uncle Jakob had died in the 1920s, and the latter's children, Robert and Alice, had left the city around that time for southern Germany. The remaining members of Einstein's extended family in Berlin were Albert and Elsa, along with her daughters, Ilse and Margot, and their husbands. Albert Einstein saw Berlin for the last time in December 1932, when he and his wife embarked on a research tour in America. Margot and Ilse followed their parents in the spring and fall, respectively, of 1933, after they had cleared out the apartment on Haberlandstrasse.

BIBLIOGRAPHY

PRIMARY SOURCES

1. *The Collected Papers of Albert Einstein*. Edited by J. Stachel, R. Schulmann, D. Kormos-Buchwald, et al. Translation series, vols. 1–13. Princeton, NJ, 1987–2012.
2. *Einsteins Annus Mirabilis: Fünf Schriften, die die Welt der Physik revolutionierten*. Edited by J. Stachel. Reinbeck, 2001.
3. *Einstein's Annalen Papers: The Complete Collection, 1901–1922*. Edited by J. Renn. Berlin, 2005.
4. *Einsteins Relativitätstheorie: Die grundlegenden Arbeiten zur Relativitätstheorie*. Edited by K. von Meyenn. Braunschweig, 1990.
5. Einstein, A. *Mein Weltbild*. Edited by C. Seelig. Zurich, 1984. (See *The World as I See It*, Zurich, 1953, or in modified translation, *Ideas and Opinions*, London, 1956.)
6. Einstein, A. *Out of My Later Years*. New York, 1950.
7. Einstein, A. "Autobiographical Notes." In P. Schilpp, ed., *Albert Einstein, Philosopher-Scientist*. La Salle, 1979.
8. Einstein, A. *Über den Frieden*. Edited by O. Nathan and H. Norden. Bern, 1975. (Translation of *Einstein on Peace*, New York, 1960.)
9. Einstein, A., and L. Infeld. *The Evolution of Physics: The Growth of Ideas from Early Concepts to Relativity and Quanta*. Cambridge, 1938.
10. Einstein, A. "Dem Gedächtnis Max Plancks." *Angewandte Chemie* 61 (1948): 113.
11. Einstein, A. "Verehrte An- und Abwesende!" Original Recordings, 1921–1951. 2 CDs. Cologne, 2003.
12. Albert Einstein Archives. The Jewish National and University Library, Hebrew University of Jerusalem, preserves the scientific papers of Albert Einstein including most of his correspondence. Many documents are available on the Internet at www.alberteinstein.info.

CORRESPONDENCE AND DOCUMENT COLLECTIONS

13. *The Quotable Einstein*. Edited by A. Calaprice. Princeton, NJ, 1996.
14. *Albert Einstein: Briefe. Aus dem Nachlaß*. Edited by H. Dukas and B. Hoffmann. Zurich, 1981.

15. Einstein, A., and A. Sommerfeld. *Briefwechsel*. Edited with commentary by A. Hermann. Basel 1968.
16. Einstein, A., and M. Born. *The Born-Einstein Letters: Correspondence between Albert Einstein and Max and Hedwig Born from 1916 to 1955, with commentaries by Max Born*. Translated by I. Born. New York, 1971.
17. Einstein, A. *Letters to Solovine*. With an introduction by M. Solovine. New York, 1987.
18. Einstein, A. and M. Besso. *Correspondence, 1903–1955*. Edited by P. Speziali. Paris, 1972.
19. Einstein, A. and M. Marić. *The Love Letters*. Princeton, NJ, 1992.
20. Seelic, C., ed. *Helle Zeit—Dunkle Zeit*. Zurich, 1956.
21. Kirsten, C., and H.-J. Treder, eds. *Albert Einstein in Berlin, 1913–1933*. 2 vols. Berlin, 1979.
22. Grüning, M. *Ein Haus für Albert Einstein: Erinnerungen, Briefe, Dokumente*. Berlin, 1990.
23. Renn, J., ed. *Albert Einstein, Ingenieur des Universums*. Dokumente seines Lebens. Berlin, 2005.
24. Aichelburg, P. C., and R. U. Sexl, eds. *Albert Einstein: His Influence on Physics, Philosophy and Politics*. Braunschweig, 1979.
25. Fölsing, A. *Albert Einstein: A Biography*. New York, 1997.
26. Frank, P. *Einstein: Sein Leben und seine Zeit*. Braunschweig, 1979. (Originally published Munich, 1949; translated as *Einstein: His Life and Times*, New York, 1947.)
27. Hermann, A. *Einstein: Der Weltweise und sein Jahrhundert*. Munich, 1994.
28. Herneck, F. *Einstein und sein Weltbild*. Berlin, 1976.
29. Highfield, R., and P. Carter. *The Private Lives of Albert Einstein*. London, 1993.
30. Hoffmann, D., and R. Schulmann. *Albert Einstein (1879–1955)*. Berlin, 2005.
31. Lohmeier, D., and B. Schell. *Einstein, Anschütz und der Kieler Kreiselkompaß: Der Briefwechsel zwischen Albert Einstein und Hermann Anschütz-Kaempfe und andere Dokumente*. Heide, 1992.
32. Neffe, J. *Einstein: Eine Biographie*. Hamburg, 2005.
33. Pais, A. *Subtle Is the Lord: The Science and the Life of Albert Einstein*. Oxford, 1982.
34. Pais, A. *Einstein Lived Here*. Oxford, 1994.
35. Seelig, C. *Albert Einstein: Leben und Werk eines Genies unserer Zeit*. Zurich, 1960.
36. Stachel, J. *Einstein from "B" to "Z."* Basel, 2002.

SPECIALIZED ANALYSES AND LITERATURE

37. Britzke, E. "Einstein in Caputh." In J. Renn, ed., *Albert Einstein, Ingenieur des Universums: Hundert Autoren für Einstein*, 272–280. Berlin, 2005.
38. Bucky, Peter A. *Der private Albert Einstein*. Düsseldorf, 1991. (Translated as *The Private Albert Einstein*, Kansas City, 1992.)
39. Castagnetti, G., H. Goenner, J. Renn, T. Sauer, and B. Scheideler. *Foundations in Disarray: Essays on Einstein's Science and Politics in the Berlin Years*. Preprint no. 63 of the Max Planck Institute for History of Science. Berlin, 1997.

40. Castagnetti, G., and H. Goenner. *Einstein and the Kaiser Wilhelm Institute for Physics (1917–1922): Institutional Aims and Scientific Results.* Preprint no. 261 of the Max Planck Institute for History of Science. Berlin, 2004.
41. Dannen, G. "Die Einstein-Szilard-Kühlschränke." *Spektrum der Wissenschaft*, June 1997, 94–100.
42. Dirks, C., and H. Simon, eds. *Relativ jüdisch: Albert Einstein, Jude, Zionist, Nonkonformist.* Berlin, 2005.
43. Franke, K. *Moritz Katzenstein: Bedeutender Berliner Chirurg-langjähriger Freund von Albert Einstein.* Berlin, 2005.
44. Goenner, H., and G. Castagnetti. "Albert Einstein as Pacifist and Democrat during World War I." *Science in Context* 9 (1996): 325–386.
45. Goenner, H. *Einstein in Berlin.* Munich, 2005.
46. Graff, W. "Albert Einstein als Erfinder in den Jahren 1907 bis 1933." Ph.D. diss., University of Stuttgart, 2005. Available at European Cultural Heritage Online, http://echo.mpiwg-berlin.mpg.de/ECHOdocuView?mode=imagepath&url=/mpiwg/online/permanent/library/AT78UEHC/pageimg.
47. Grundmann, S. *The Einstein Dossiers: Science and Politics—Einstein's Berlin Period.* Heidelberg, 2005. (Translation of *Einsteins Akte: Wissenschaft und Politik—Einsteins Berliner Zeit*, 2nd ed., Heidelberg, 2004.)
48. Hentschel, K. *Interpretationen und Fehlinterpretationen der speziellen und der allgemeinen Relativitätstheorie durch Zeitgenossen Albert Einsteins.* Basel, 1990.
49. Hentschel, K. *The Einstein Tower: An Intertexture of Dynamic Construction, Relativity Theory, and Astronomy.* Stanford, CA, 1997. (Expanded translation of *Der Einstein Turm*, Berlin, 1992.)
50. Hermann, A. *Der Weg ins Atomzeitalter: Physik wird Weltgeschichte.* Munich, 1979.
51. Herneck, F. *Einstein privat: Herta B. erinnert sich an die Jahre 1927 bis 1933.* Berlin, 1979.
52. Herrmann, D. "Einstein, Archenhold und die Popularisierung der Naturwissenschaften." In *Astronomiegeschichte. Ausgewählte Beiträge zur Entwicklung der Himmelskunde*, 327–333. Berlin, 2004.
53. Hobohm, M. "Einstein und Gumbel." *Die Menschenrechte* 6 (1931): 122–124. "Die Hochschulreaktion." *Die Menschenrechte* 6 (1931): 99–111.
54. Hoffmann, D. "Einstein und die Physikalisch-Technische Reichsanstalt." *ITW-Kolloquien* 21 (1980): 90–102. (Shortened version published as "The Einstein Case," *PTB-Mitteilungen* 122, no. 2 [2012]: 22–25.)
55. Hoffmann, D. "Physics in Berlin." In J. S. Rigden and R. H. Stuewer, eds., *The Physical Tourist: A Science Guide for the Traveler*, 81–110. Basel, 2008.
56. Hoffmann, D. "Ein Experimentalphysiker als antitheoretischer Sammler." In A. te Heesen, ed., *Cut and Paste um 1900: Der Zeitungsausschnitt in den Wissenschaften. Kaleidoskopien* 4 (2002): 70–81.
57. Hoffmann, D. "Einsteins politische Akte." *Physik in unserer Zeit* 35 (2004): 64–69.
58. Hoffmann, D. "Albert Einstein und die Friedrich-Wilhelms-Universität." *Impulse der Physik: Zeitschrift der Vereinigung der Freunde und Förderer des Instituts für Physik der Humboldt-Universität zu Berlin*, October 2004, 17–22.

59. Hoffmann, D. "Albert Einstein und die Deutsche Physikalische Gesellschaft." *Physik Journal* 4 (2005): 85–90.
60. Hoffmann, D. "Auf Einsteins Spuren." Series I–VIII. *Physik in unserer Zeit* 35, nos. 2–6; 36, nos. 1–4 (2005).
61a. Hoffmann, D. "Einstein's Berlin." In J. Renn, ed., *Albert Einstein, Chief Engineer of the Universe: One Hundred Authors for Einstein*, 260–265. Berlin, 2005.
61b. Illy, J. *Albert Meets America: How Journalists Treated Genius during Einstein's 1921 Travels.* Baltimore, 2006.
61c. Illy, J. *The Practical Einstein: Experiments, Patents, Inventions.* Baltimore, 2012.
62. Jerome, F. *The Einstein File: J. Edgar Hoover's War against the World's Most Famous Scientist.* New York, 2002.
63. Kant, H. "Albert Einstein, Max von Laue, Peter Debye und das Kaiser-Wilhelm-Institut für Physik in Berlin (1917–1939)." In B. vom Brocke and H. Laitko, eds., *Die Kaiser-Wilhelm/Max-Planck-Gesellschaft und ihre Institute*, 227–243. Berlin, 1996.
64. Kirsten, C., and H.-G. Körber, eds. *Physiker über Physiker.* Vols. 1 and 2. Berlin, 1975 and 1979.
65. Kohn, W. "Erinnerungen an Albert Einstein." *Freitag*, 22 April 2005, supplement, iii.
66. Laitko, H., et al., eds. *Wissenschaft in Berlin.* Berlin, 1987.
67. Levenson, J. *Einstein in Berlin.* New York, 2003.
68. Melcher, H. "Einsteins Patente." *Spektrum*, 1978, issue 9, pp. 23–26.
69. Mechler, W.-D. "Einsteins Wohnungen in Berlin." In J. Renn, ed., *Albert Einstein, Chief Engineer of the Universe: One Hundred Authors for Einstein*, 266–271. Berlin, 2005.
70. Plesch, J. *János: The Story of a Doctor.* London, 1947. (Translated into German as *János erzählt von Berlin*, Munich, 1955.)
71. Renn, J., G. Castagnetti, and P. Damerow. "Albert Einstein: Alte und neue Kontexte in Berlin." In J. Kocka, ed., *Die Königlich Preußische Akademie der Wissenschaften zu Berlin im Kaiserreich*, 333–354. Berlin, 1999.
72a. Rompe, R. ". . . eine faszinierende Persönlichkeit." *Wissenschaft und Fortschritt* 29 (1979): 47–48.
72b. Rosenkranz, Z. *Einstein before Israel: Zionist or Iconoclast?* Princeton, NJ, 2011.
72c. Rowe, D. E., and R. Schulmann, eds. *Einstein on Politics: His Private Thoughts and Public Stands on Nationalism, Zionism, War, Peace, and the Bomb.* Princeton, NJ, 2007.
73. Scheideler, B. "The Scientist as Moral Authority: Albert Einstein between Elitism and Democracy, 1914–1933." *Historical Studies in the Physical Sciences* 32 (2002): 319–346.
74. Schulz, F., and E. Schwarz. *Einstein in Ahrenshoop.* Kückenshagen, 1995.
75. Seghers, A. "Ich fuhr zu ihm nach Caputh." *Spektrum*, 1978, issue 12, pp. 12, 15–16.
76. Stern, F. "Freunde im Widerspruch: Albert Einstein und Fritz Haber." In T. Karlauf, ed., *Deutsche Freunde*, 222–254. Berlin, 1995.
77. Strauch, D. *Einstein in Caputh: Die Geschichte eines Sommerhauses.* Berlin, 2001.

GENERAL LITERATURE AND OTHER QUOTED SOURCES

78. Brocke, B. vom, ed. *Wissenschaftsgeschichte und Wissenschaftspolitik im Industriezeitalter: Das "System Althoff" in historischer Perspektive.* Hildesheim, 1991.
79. Chaplin, C. *Die Geschichte meines Lebens.* Stuttgart, 1964. (Translation of *My Autobiography*, London, 1964.)
80. Elias, R. *Die Gesellschaft der Freunde des neuen Russland.* Cologne, 1985.
81. Franke, O. *Erinnerungen aus zwei Welten: Randglossen zur eigenen Lebensgeschichte.* Berlin, 1954.
82. Goll, C. *Ich verzeihe keinem: Eine literarische Chronique scandaleuse.* Berlin, 1980.
83. Grüning, M. *Der Wachsmann-Report: Auskünfte eines Architekten.* Berlin, 1985.
84. Hannak, J. *Emanuel Lasker: Biographie eines Schachweltmeisters.* Berlin, 1952.
85. Hentschel, K., ed. *Physics and National Socialism: An Anthology of Primary Sources.* Basel, 1996.
86. Kisch, E. E. *Gesammelte Werke.* Vol. 10. Berlin, 1985.
87. Kessler, H. *Tagebücher, 1918–1937.* Frankfurt am Main, 1971.
88. Lehmann-Russbüldt, O. *Der Kampf der deutschen Liga für Menschenrechte für den Weltfrieden, 1914–1927.* Berlin, 1927.
89. Meinecke, F. *Strassburg, Freiburg, Berlin, 1901–1919: Erinnerungen.* Stuttgart, 1919.
90. Planck, M. Speech at the plenary meeting of the Academy from 14 November 1918. *Sitzungsberichte der Preußischen Akademie der Wissenschaften* (1918), 992–993.
91. Rolland, R. *Das Gewissen Europas: Tagebuch der Kriegsjahre, 1914–1919.* Vol. 1. Berlin, 1963.
92. Sommerfeld, A. *Wissenschaftlicher Briefwechsel.* Vols. 1 and 2. Edited by M. Eckert and K. Märker. Diepholz, 2000, 2004.
93. Ungern-Sternberg, J., and W. Ungern-Sternberg. *Der Aufruf "An die Kulturwelt": Das Manifest der 93 und die Anfänge der Kriegspropaganda im Ersten Weltkrieg.* Stuttgart, 1996.
94. Wendel, G. *Die Kaiser-Wilhelm-Gesellschaft, 1911–1914: Zur Anatomie einer imperialistischen Forschungsgesellschaft.* Berlin, 1975.
95. Willstätter, W. *Aus meinem Leben: Von Arbeit, Muße und Freunden.* Weinheim, 1949.
96. Archiv zur Geschichte der Max Planck Gesellschaft, Berlin. Papers of Otto Hahn and Max von Laue.
97. Archiv der Eidgenössisch-Technischen Hochschule, Zurich. Papers of Carl Seelig.
98. Rijks Archjiv Haarlem. Papers of H. A. Lorentz.
99. Staatsbibliothek Berlin, Manuscripts department. Papers of H. A. Krüss.

ILLUSTRATION CREDITS

My sincere thanks go to the staffs of the archives and to all the individuals who very kindly provided us with images and gave their permission to print the images. Despite careful research, in some cases we were unable to identify or locate the copyright holders.

Archenhold Observatory, Berlin: Figs. 2.23, 2.24, 2.27

Berlin-Brandenburg Academy of Science and Humanities Archives: Figs. 1.12, 2.1, 2.2

Britzke, Erika (Wilhelmshorst): Fig. 1.16

Centrum Judaicum, Berlin: Figs. 3.12, 3.13, 3.14, 3.15

Deutsches Technik Museum Archives, Berlin: Fig. 2.18

Dreike, Ralph (Mountain View, CA): Fig. 4.12

German Historical Museum, Berlin: Figs. 1.10, 1.15

Graff, Wolfgang (Ludwigsburg): Fig. 2.20

Gruss-Castel, Brigitte (Düsseldorf): Fig. 4.7

Hebrew University, Jerusalem: Figs. 1.3, 1.6, 1.14, 2.3, 2.19, 3.8

Jäger, Friedrich (Potsdam): Figs. 2.26, 2.28

Kleinert, Andreas (Halle/Saale): Fig. 3.7

Landesarchiv Berlin: Figs. 1.2, 1.4, 2.4, 2.9, 3.6, 3.9, 3.16, 3.17, 4.9

Leo Baeck Institute, New York City: Figs. 1.5, 4.8

Lotte Jacobi Collection, University of New Hampshire: Fig. 4.13

Matthaei, Christian (Frankfurt/Main): Fig. 4.1

Max Planck Society Archives: Figs. 0.1, 2.8, 2.11, 2.12, 2.13, 2.14, 4.2, 4.3

Mechler, Wolf-Dieter (Hannover): Fig. 1.1

Museum Berlin-Charlottenburg/Wilmersdorf: Figs. 3.1, 3.2

Museum Berlin-Zehlendorf: 4.11

Niels Bohr Library, American Institute of Physics, College Park, MD: Fig. 2.22

The Pasternak Trust, Oxford: Fig. 3.10

Illustration Credits

Physikalisch-Technische Bundesanstalt, Brunswick: Figs. 2.15, 2.16, 2.17

Stiftung Preußischer Kulturbesitz Picture Archives, Berlin: Fig. 1.9

Ullstein-Bilderdienst, Berlin: 1.8, 3.3, 3.4, 4.4

Photo collection of the author: Figs. 1.13, 1.7, 1.11, 2.5, 2.6, 2.7, 2.10, 2.14, 2.21, 2.25, 2.29, 3.5, 3.11, 4.5, 4.6, 4.10, 4.14, 4.15

NAME INDEX

Abbe, Franz, 8
Althoff, Friedrich, 48, 49, 50, 53, 54
Ampere, Andre Marie, 68, 70
Amundsen, Roald, 83
Andersen, Hans Christian, 133
Anschütz-Kaempfe, Hermann, 72
Archenhold, Friedrich Simon, 83, 84
Arco, Count Georg von, 97, 119, 132

Bach, Johann Sebastian, 126
Baeck, Leo, 12
Baeyer, Otto von, 3
Barbusse, Henri, 121
Baron, Erich, 117
Becker, Carl Heinrich, 52
Beethoven, Ludwig van, 114, 115, 139
Behring, Emil von, 50
Bergmann, Hugo, 122
Berliner, Arnold, 132
Besso, Michele, 11, 32, 64, 69, 85, 106, 132, 147
Bismarck, Otto von, 50
Blumenfeld, Kurt, 131
Bohr, Niels, 64
Boltzmann, Ludwig, 38
Born, Hedwig (Hedi), 23, 113, 122, 133
Born, Max, 71, 106, 108, 113, 116
Bose, Satyendra, 3
Bothe, Walther, 67, 71
Brecht, Bertolt, xii, 85
Breuer, Marcel, 22
Brugsch, Theodor, 147
Bucky, Gustav, 132

Chaplin, Charlie, 13
Colin, Paul, 101
Copernicus, Nicolaus, 138

Debye, Peter, 65
Dorn, Ernst, 69
Dukas, Helen, 23, 64, 125

Ebert, Friedrich, 107
Eddington, Arthur Stanley, 33, 86, 88
Edison, Thomas Alva, 128
Ehrenfest, Paul, 23, 96, 100, 125, 139
Ehrmann, Rudolf, 132, 148
Einstein, Alice, 157
Einstein, Bernhard Caesar, 23, 24
Einstein, Eduard, 7, 23, 59
Einstein, Elsa (Löwenthal), 2, 6, 9, 10, 11, 12, 13, 14, 15, 17, 18, 19, 22, 26, 42, 48, 53, 59, 137, 149, 150, 151, 155, 156, 157
Einstein, Fanny, 12, 156
Einstein, Hans Albert, 7, 8, 23, 24, 59, 77
Einstein, Ilse (née Löwenthal), 12, 13, 23, 64, 157
Einstein, Maja (Winteler), 23, 26, 104, 155
Einstein, Margot (née Löwenthal), 12, 13, 23, 120, 157
Einstein, Mileva (née Marič), 6, 7, 10, 53, 59, 149, 157
Einstein, Pauline (née Koch), 5, 105, 150, 156, 157
Einstein, Robert, 157
Einstein, Rudolf, 12, 155, 156
Eisfelder, Otto, 12
Eisler, Hanns, 85

Name Index

Eisner, Olga, 74
Esser, Bruno, 14

Faraday, Michael, 15
Fermi, Enrico, 81
Fersman, Alexander J., 119
Finlay-Freundlich, Erwin, 52, 86, 87, 88, 89, 90, 91, 93, 94
Fischer, Berman, 146
Fischer, Samuel, 131
Franck, James, 3, 42
Franck, Ingrid, 3
Frank, Philipp, 13, 16, 39
Franke, Otto, 31
Friedrich, Walter, 134
Friedrich Willhelm III (King of Prussia), 38, 39

Galilei, Galileo, xi
Gehrcke, Ernst, 111, 112, 113
Geiger, Hans, 67, 71
Goebbels, Joseph, 130
Goldscheid, Rudolf, 98
Goldschmidt, Rudolf, 72, 73, 74, 75, 144
Goldstein, Eugen, 42
Goll, Claire, 149
Grotrian, Walter, 3
Grünberg, Josef, 48, 132
Gumbel, Emil Julius, 102, 103

Haas, Wander Johannes de, 45, 68, 69
Haber, Clara (née Immerwahr), 5, 52
Haber, Fritz, 3, 5, 7, 12, 44, 47, 52, 55, 56, 57, 58, 59, 60, 61, 62, 96, 132, 140, 148
Haenisch, Konrad, 89, 114
Hahn, Otto, 3, 139
Hannak, Jacques, 144
Harm, Adolf, 25
Harnack, Adolf von, 55, 61
Hauptmann, Gerhart, 13, 96, 131
Hellpach, Willy, 122
Helmholtz, Hermann von, 42, 65
Herneck, Friedrich, 152, 153
Hertz, Gustav, 3, 42
Hertz, Heinrich, 42, 129
Hettner, Gerhard, 38
Heymann, Ernst, 35
Hobohm, Martin, 103
Hoff, Jacobus Henricus van't, 29

Hurwitz, Adolf, 31

Ihne, Friedrich von, 29
Infeld, Leopold, 82
Isenstein, Harald, 93, 94

Jadlowker, Hermann, 126, 127
Juliusburger, Otto, 132

Katzenellenbogen, Estella, 154
Katzenstein, Moritz, 132, 144, 145, 146, 150, 151
Kaufmann, Walter, 137
Kerr, Alfred, 12
Kessler, Count Harry, 13, 101, 131
Kestenberg, Leo, 98
Kisch, Egon Erwin, 40
Kleiber, Erich, 132
Knipping, Paul, 134
Koch, Jakob, 5, 156, 157
Koch, Robert, 50
Koesters, Werner, 71
Kohlschütter, Ernst, 52
Kohn, Walter, vii, viii, ix
Konen, Hermann, 46
Koppel, Leopold, 1, 13, 58, 59, 131
Kraus, Friedrich, 147
Kreisler, Fritz, 132, 148
Krüss, Hugo Andres, 50, 51, 52, 53, 54

Landsberger, Artur, 12
Lasker, Eduard, 144
Lasker, Emanuel, 12, 131, 141, 142, 143, 144
Lasker-Schüler, Else, 12
Laub, Jakob, 1, 28
Laue, Max von, 23, 33, 34, 35, 38, 39, 43, 44, 45, 47, 64, 76, 112, 113, 132, 133, 134, 135, 136, 137, 140
Lebach, Margarete, 153
Lebedew, Peter, 42
Lehmann-Russbüldt, Otto, 98, 101
Lenard, Philipp, 113, 114
Lewandowski, Alfred, 126, 127
Lieben, Robert von, 129
Liebermann, Max, 23, 131
Linde, Carl von, 70
Löbe, Paul, 117
Lowell, Percival, 83
Löwenthal, Elsa. See Einstein, Elsa

Löwenthal, Ilse. *See* Einstein, Ilse
Löwenthal, Margot. *See* Einstein, Margot
Löwenthal, Max, 155, 157
Ludendorff, Hans, 93
Lummer, Otto, 42
Lunatsharski, Anatoli, 120

Mann, Heinrich, 23
Mann, Thomas, 118
Marianoff, Dmitri, 120, 150
Maxwell, James Clerk, 15, 129
Mayer, Walther, 23
Meinecke, Friedrich, 33
Meissner, Walther, 67
Meitner, Lise, 3, 100, 139
Mendel, Bruno, vii, 152
Mendel, Toni, vii, ix, 151, 152, 153
Mendelsohn, Erich, 91, 131
Mendelssohn, Francesco von, 14
Michelson, Albert Abraham, 42
Millikan, Robert Andrew, 136
Moissi, Alexander, 114
Mössbauer, Rudolf, 93
Moszkowski, Alexander, 142
Mühsam, Hans, 132, 150

Nansen, Fridtjof, 83
Nernst, Walther, 8, 18, 29, 42, 57, 58, 61, 62, 70, 96, 113, 135, 136
Neumann, Betty, 150
Newton, Isaac, 40, 87
Nicolai, Georg Friedrich, 97, 132
Nikleniewicz, Johann, 5

Oersted, Hans Christian, 129
Oppenheim, Max, 131
Orlik, Emil, 131

Pasternak, Boris, 117
Pasternak, Leonid, 117, 118, 131
Pasternak-Slater, Lydia, 117
Perlin, Frida, 102
Planck, Erwin, 139
Planck, Karl, 138
Planck, Marga, 45, 138, 141
Planck, Max, 12, 29, 32, 35, 45, 46, 47, 48, 50, 57, 58, 61, 67, 70, 87, 88, 96, 100, 132, 133, 134, 135, 136, 137, 138, 139, 140, 141

Plesch, Janos, 46, 147, 148, 149
Poelzig, Hans, 118, 132
Pringsheim, Peter, 3, 42
Pulsack, Elise, 12
Pupin, Michael, 42

Radványi, László, 84
Ramsauer, Carl, 78
Rantzau, Count, 148
Rathenau, Walther, 13, 41, 131
Regener, Erich, 42
Reichenbach, Hans, 38
Reinhardt, Max, 96, 114
Rolland, Romain, 98, 99
Rompe, Robert, 38, 44
Röntgen, Wilhelm Conrad, 29, 70
Roosevelt, Franklin D., 80
Rosenberg, Artur, 103
Roskin, Janot S., 127
Rubens, Heinrich, 58, 61, 62, 134

Scheel, Karl, 45
Schlick, Moritz, 138
Schmidt-Ott, Friedrich, 50, 51, 52
Schnabel, Arthur, 148
Schopenhauer, Arthur, 15
Schrödinger, Erwin, 23, 38, 136
Schubert, Franz, 139
Schücking, Walther, 98
Schwarz, Joseph, 132
Schwarzschild, Karl, 89, 90
Seeberg, Reinhold, 106
Seelig, Carl, 132
Seghers, Anna, 23, 84
Siemens, Werner von, 66
Silberstein, L., 140
Simon, Hugo, 131
Slevogt, Max, 48, 148
Solovine, Maurice, 125
Sommerfeld, Arnold, 33, 35, 45, 65, 86, 112
Spinoza, Baruch, 126
Sponer, Hertha, 3
Stalin, Joseph W., 121
Stern, Otto, 31
Stock, Franz, 61
Stresemann, Gustav, 131
Stroux, Johannes, 35
Struve, Karl Hermann, 87

Szilard, Leo, 76, 77, 78, 79, 80, 81

Tagore, Rabindranath, vi, 23, 24, 153
Tobler, Adolf, 155
Trotzki, Leo, 121

Wachsmann, Konrad, 13, 16, 21, 22, 23, 134, 150
Waldow, Hertha, 151
Warburg, Emil, 42, 45, 58, 61, 62, 66, 67, 68, 132
Warburg, Otto, 67
Wegener, Alfred, 83
Weizmann, Chaim, 125

Wertheimer, Max, 106
Westphal, Wilhelm, 3, 38
Weyland, Paul, 110, 111, 112, 113, 114
Wien, Wilhelm, 42, 44
Wigner, Eugene, 80
Wilhelm II (Emperor of Germany), 33, 54, 69
Willstätter, Karl, 30

Zangger, Heinrich, 8, 9, 31, 32, 66, 148
Zepke, Arthur, 126, 127
Zweig, Arnold, 23
Zweig, Stefan, 114

LOCATION AND STREET INDEX

Amsterdam, 101

Baltic Sea, 13
Berlin
  Abbestrasse, 66
  Ahornallee, 144, 145
  Albertinenstrasse, 133, 134, 135, 136
  Alt-Treptow, 81
  Am Reichstagsufer, 42, 43, 46
  Am Sandwerder, 152, 153
  Anhalter Bahnhof, 108
  Aroser Allee, 78
  Aschaffenburger Strasse, 12, 14, 141, 142
  Auguststrasse, 145
  Bavarian Quarter, 12
  Bayerischer Platz, 10, 141
  Bendlerstrasse, 154
  Bernburgerstrasse, 108
  Boltzmannstrasse, 65
  Boxfelde, 18
  Budapester Strasse, 148
  Burgunderweg, 18
  Charlottenburg, 66, 71, 76, 128, 144, 156
  Dahlem, 5, 54, 56, 58, 65, 79
  Dorotheenstrasse, 42
  Ehrenbergstrasse, 5, 6, 7
  Eisackstrasse, 156
  Ernst-Reuter-Platz, 66
  Faradayweg, 54
  Fehrbelliner Platz, 7, 8
  Franz-Neumann-Platz, 72
  Friedrich Karl Strasse (now Am Sandwerder), 152, 153
  Friedrichshain, 145, 146
  Friedrichstrasse, 28, 38, 42, 104
  Gatow, 147
  Giesbrechtstrasse, 156
  Grunewald, 137, 139
  Haberlandstrasse, 10, 11, 12, 14, 15, 16, 21, 62, 63, 64, 67, 80, 131, 135, 142, 150, 155, 157
  Hackescher Markt, 122
  Halensee, 137
  Hammaskjöldplatz, 128
  Hausvogteiplatz, 28, 38
  Heerstrasse, 147
  Hittorfstrasse, 7
  Holländerstrasse, 72, 78
  Kaiserdamm, 128
  Kavalierstrasse, 118
  Kladow, 19
  Konstanzer Strasse, 8
  Kreuzberg, 108
  Kurfürstenstrasse, 97
  Leipziger Strasse, 98, 115, 116, 119
  Lichterfelde West, 5, 54
  Luisenstrasse, 102
  Marxstrasse (now Eisackstrasse), 156
  Messe Nord 7ICC, 128
  Moabit, 72, 77
  Mohrenstrasse, 48
  Mommsenstrasse, 156
  Monbijouplatz, 97
  Nördlingerstrasse, 12
  Opernplatz, 55
  Oranienburger Strasse, 122, 127
  Pankow, 118
  Plänterwald, 81
  Platz der Republik, 104

Berlin (cont.)
  Potsdamer Platz, 108, 115
  Potsdamer Strasse, 29
  Reinickendorf, 72, 80
  Reuchlinstrasse, 77
  Rothenbücher Weg, 147
  Rudeloffweg, 5
  Schiffbauerdamm, 98
  Schöneberg, 10, 12, 141, 156
  Spandau, 18
  Spichernstrasse, 96, 97, 100
  Tauentzienstrasse, 97
  Theodor-Heuss-Platz, 128, 144, 147
  Thielplatz, 6
  Treptow, 28, 54, 81, 82
  Treptower Park, 81
  Uhlandstrasse, 12
  Unter den Linden, 28, 29, 30, 33, 38, 42, 48, 55, 104, 117
  Viktoriastrasse, 98
  Wangenheimstrasse, 137, 138, 139
  Wannsee, 2, 152
  Weinmeisterstrasse, 85
  Westend, 145
  Wilhelmstrasse, 48, 49, 97, 107
  Wilmersdorf, 7, 8, 12, 96
  Wilmersdorfer Strasse, 6, 156
  Wittelsbacher Strasse, 8, 9
  Zehlendorf, 133
  Zoo, 154
Brussels, 66
Budapest, 76

Cambridge, 27
Caputh, 18, 19, 21, 22, 23, 24, 25, 84, 135, 143, 147
  Am Waldrand, 20
  Potsdamer Strasse, 19, 26
  Waldstrasse, 18
Copenhagen, 65
Crimea, 88

Far East, 150
Ferch, 18
Frankfurt am Main, 133

Hamburg, 77
Hamilton, Ontario, vii

Havel, 19, 21
Hechingen, 155
Heidelberg, 102

Japan, 41
Jerusalem, xiii, 26

Lake Geneva, 99
Leipzig, 135
Leyden, 125, 139
Long Island, 80
Ludwigsfelde, 143

Moscow, 117, 118
Munich, 18, 29, 65

New York, 143, 144
  Bronx, 13
Niesky, 21

Paris, 17
Pasadena, 26, 34, 39, 92
Potsdam, 18, 19, 25, 40, 86, 89, 90, 91, 92, 93
  Albert-Einstein-Strasse, 86
  Telegraphenberg, 86, 87, 92, 93
Prague, xii, 16, 39
Princeton, vii, ix, 17

Sachsen-Anhalt, 76
Santa Barbara, California, ix
Schwielochsee, 18
Silicon Valley, ix, 27
Stanford, 27
Sumatra, 88
Switzerland, 155

Templiner See, 21
Thyrow, 143
Ticino, 5
Toronto, vii

Vevey, 99
Vienna, 150, 151

Washington, D.C., 140
Württemberg, 155

INSTITUTION INDEX

Academy of the Arts (Akademie der Künste der DDR), 103
Academy of Sciences, Royal, Prussian, or German, 2, 4, 5, 27, 28, 29, 31, 32, 33, 34, 38, 43, 54, 57, 69, 87
Academy of Sciences of the German Democratic Republic, 26
Academy of Sciences of the USSR, 120
AEG (Allgemeine Elektrizitäts), 27, 72, 73, 74, 75, 76, 77, 78, 80
Albert Einstein Donation Fund (Albert-Einstein-Spende), 90, 91
American Academy of Sciences, 140
Anti-Orloog-Rad (antiwar council), 100
Archenhold Observatory, 27, 81, 82, 83, 84
Astrophysical Observatory, Potsdam, 27, 89, 90, 91, 93

Bamang-Meguin, 76, 77
Bavarian Academy, 35
Beethoven-Hall of the Berlin Philharmonic, 114, 115
Berlin Palace, 56
Bund "Neues Vaterland" / Deutsche Liga für Meschenrechte. *See* New Fatherland League / German League of Human Rights

Carl Zeiss and Schott & Associates, 90
Charité, 147
Christoph & Umack, Niesky, 21
Citogel-Gesellschaft, 77

Demokratischer Club, 98

Deutsche Forschungsgemeinschaft (DFG). *See* German Research Association
Deutsche Gesellschaft zum Studium Osteuropas. *See* German Society for the Study of Eastern Europe
DPG (Deutsche Physikalische Gesellschaft). *See* German Physical Society

Einstein Archive, Jerusalem, xiii
Einstein Forum, Potsdam, 26
Einstein Institute, 93
Einstein Tower, 27, 86, 87, 92, 94
Electrolux, 76
exhibition grounds, 128, 129, 130

Federal Polytechnic Institute, Zurich, 66
Foreign Office, 53, 117
French Embassy, 17

German Entomology Museum, 65
German Physical Society (DPG), 27, 44, 45, 46, 48, 69
German Research Association, 65
German Society for the Study of Eastern Europe, 117
German University, Prague, 122
Gesellschaft für kulturelle Verbindung mit dem Ausland (VOKS). *See* Soviet Society for Cultural Ties Abroad

Haus des Deutschen Sports, 98
Hebrew University, Jerusalem, 26
Helmholtz Center Berlin, ix
Helmholtz Society, 65

Heraeus Foundation, ix, xiii
Hospital Friedrichshain, 145, 146
Humboldt University. *See* University of Berlin

Imperial Institute of Physics and Technology (PTR), 27, 45, 66, 67, 68, 69, 70, 71, 72, 132
Institute for Theoretical Physics, Santa Barbara, vii
Institut of Advanced Study (Princeton), vii
International Workers Relief (Internationale Arbeiterhilfe), 116

Jewish Community Berlin (Jüdische Gemeinde Berlin), 126
Jewish Student Association (Verband jüdischer Studenten), 115
Jewish Worker's Bureau (Jüdische Arbeitsamt), 124

Kaiser Wilhelm Institute of Chemistry, 30, 59
Kaiser Wilhelm Institute of Physical Chemistry and Electrochemistry, 54, 55, 56, 59
Kaiser Wilhelm Institute of Physics, 2, 3, 12, 51, 56, 58, 61, 62, 63, 65
Kaiser Wilhelm Society, 2, 5, 27, 43, 51, 54, 55, 56, 58, 59, 61, 64, 91, 135, 140
Kopple Foundation, 58

Langenbeck-Virchowhaus, 102
League of Nations, 50

MASCH (Marxist School of Workers), 23, 84, 85
Mathematisch-Physikalische Arbeitsgemeinschaft (Mapha), 41
Max Planck Institute for the History of Science, ix, xiii
Max Planck Society, 57
Messegelände *See* exhibition grounds
Metropol Theater, 40
MIT (Massachusetts Institute of Technology), 27
Mount Wilson Observatory, 92

New Fatherland League / German League of Human Rights, 96, 97, 98, 100, 101, 102, 104, 116, 117, 151

New Synagogue, 122, 123, 124, 125, 126, 127

Office of Intellectual Property (Patentamt), 133
Ostwerke, 154

palace. *See* Berlin Palace
parliament building (Reichstag), 98, 104, 105, 106, 107
People's Chamber of the East German Republic (Volkskammer der DDR), 103
Philharmonic Hall (Philharmonie), 108, 109, 110, 111, 112, 113, 114
Physical Institute of the University of Berlin, 42, 43, 44, 46, 47
Physical Society in Berlin (Physikalische Gesellschaft zu Berlin), ix, 44
Physikalisch-Technische Reichsanstalt/Bundesanstalt (PTR/PTB). *See* Imperial Institute of Physics and Technology
Polytechnic Institute, Berlin-Charlottenburg (Technische Hochschule), 76, 77
Prussian Ministry of Culture / Ministry of Religious, Educational, and Medical Affairs, 48, 49, 50, 53, 54, 56, 57, 58, 89, 114
Prussian Ministry of Finance, 59
Prussian Ministry of Interior, 69, 71
Prussian Upper House (Herrenhaus), 98, 116, 119

Radio Fair (Funkausstellung), 128, 130
Reich's Chancellor's Palace (Reichskanzlerpalais), 107, 139
Reichstag. *See* parliament building
Rockefeller Foundation, 65
Russian Embassy, Berlin, 117, 118

Society of Friends of the New Russia (Gesellschaft der Freunde des Neuen Russland), 115, 116, 117, 118, 119, 120, 121
Society of German Scientists and Medical Doctors (Gesellschaft deutscher Naturforscher und Ärzte), 135
Solvay conference, 1, 57, 66
Soviet Chamber of Commerce, Berlin, 120
Soviet Society for Cultural Ties Abroad (VOKS), 118
State Library, Royal Library (Königlich preus-

sische Staatsbibliothek), 29, 30, 39, 50, 53, 55

University of Berlin, 2, 5, 31, 38, 39, 50, 64, 76, 86, 146, 147, 155
University of Frankfurt a.M., 133
University of Heidelberg, 102

Villa Lemm, 147, 148
Volkshochschule Berlin, 85

Wissenschaftskolleg, Berlin, 59
World Exhibition, St. Louis, 50

About the Author

Dieter Hoffmann studied physics at Humboldt University in Berlin from 1967 to 1972, earning his doctorate there in 1976 and qualifying for academic teaching (habilitation) in 1989 in the history of science. He worked as a historian of science at the Academy of Sciences of the German Democratic Republic from 1976 to 1991, after which he had various jobs and held a Humboldt-Stiftung stipend for studies at Stuttgart, Harvard, and Cambridge. Since 1996 he has been a senior fellow at the Max Planck Institute for the History of Science in Berlin while also serving as adjunct professor at Humboldt University. In 2001 he was elected a member of the International Academy of the History of Science and in 2010 he was elected a member of the Leopoldina, National Academy of Sciences. Professor Hoffmann has published numerous works in the history of physics and the history of science of the nineteenth and twentieth centuries that testify to his profound knowledge of the life and work of Albert Einstein.